Master Plumber's Licensing Exam Guide

Master Plumber's Licensing Exam Guide

R. Dodge Woodson

McGraw-Hill
New York San Francisco Washington, D.C. Auckland Bogotá
Caracas Lisbon London Madrid Mexico City Milan
Montreal New Delhi San Juan Singapore
Sydney Tokyo Toronto

Published by McGraw-Hill, Inc.

pbk 1 2 3 4 5 6 7 8 9 0 FGR/FGR 9 9 8 7 6 5 4
hc 1 2 3 4 5 6 7 8 9 0 FGR/FGR 9 9 8 7 6 5 4

Library of Congress Cataloging-in-Publication Data
Woodson, R. Dodge (Roger Dodge), 1955–
Master plumber's licensing exam guide / by R. Dodge Woodson.
p. cm.
Includes index.
ISBN 0-07-071786-9 (p) ISBN 0-07-071785-0 (h)
1. Plumbing—Examinations, questions, etc. I. Title.
TH6128.W593 1994
696.1'076—dc20 94-30233
CIP

Acquisitions editor: April Nolan
Editorial team: Joanne M. Slike, Executive Editor
Lori Flaherty, Supervising Editor
Barbara Minich, Book Editor
Production team: Katherine G. Brown, Director
Rhonda E. Baker, Coding
Donna K. Harlacher, Coding
Lisa M. Mellott, Desktop Operator
Nancy K. Mickley, Proofreading
Joann Woy, Indexer 0717869
Designer: Jaclyn J. Boone WK2

To my wonderful daughter, Afton Amber Woodson, and Kimberley, the best wife a man could hope for.

Acknowledgments

I would like to thank Dawn Gerald-Hall for her administrative support in putting this project together. Dawn's efforts have allowed me to put these pages into your hands faster.

I would also like to thank my parents, Maralou and Woody, for their many years of help in all of my endeavors.

Contents

Introduction

Do you want a fulfilling career, the growth potential to own your own lucrative business, and the freedom to call your own shots? Are you interested in making more money in a single day than many people make in a week? If you are, you're reading the right book, because licensed plumbers can do all of this and more.

If you are hoping to become a master plumber, you're holding the key to your success. This book might prove to be your ticket to a better career, more money, a happier life, and a true sense of satisfaction. Can a single book do all of that for you? Probably not by itself, but this book has the power, when combined with your efforts, to prepare you for passing the licensing exam required to receive a master plumber's license.

I have combined my 20 years of plumbing knowledge, my experience as a teacher, and a reader-friendly writing style to make this book an invaluable tool for anyone wishing to become a licensed master plumber.

Learning the plumbing code can be difficult. If you don't have a master plumber supporting your study efforts, it is easy to become confused, frustrated, and concerned about your ability to comprehend the cryptic language so often used in the code. I recognize the problems my students and employees often have interpreting what the plumbing code means. *Master Plumbers's Licensing Exam Guide* takes the confusing code and transforms it into plain English. With this book in your hand, I will become the master plumber who supports your efforts.

Not everyone who wants to become a licensed master plumber has the luxury of working for a plumbing company that encourages the development of master plumbers. Some companies are afraid of their journeyman plumbers moving up to the level of master plumbers. They feel it is a threat to their business if their employees are master plumbers and have the ability to become direct competition.

Many people enter a vocational school and work toward earning their plumbing licenses academically. This process works, but it lacks the real-world experience that people in the field gain from being exposed to working plumbers. Since I am both a working plumber and a teacher, I am able to bring more to the table than the average teacher or plumber. My combined experience and knowledge allows me to help you in ways that others can't.

Will this book change your life? It might. Once you obtain your master's license, you will be able to open your own plumbing business. Whether you work as an individual plumber or hire dozens of plumbers to work for you (I have done both), you will have the opportunity to carve your own outlook into the face of the future.

Whether you are a plumbing apprentice who wants to move up the ladder of success or whether you are a bank teller who is tired of dressing up every day, this book can give you a big boost towards becoming a licensed master plumber. Everyone has heard the jokes about how much money plumbers make, right? Well, guess what, the jokes are not too far from the truth. Seasoned master plumbers can make upwards of $60 an hour, and any licensed plumber can earn a better-than-average income.

What will you find in the pages of this book? You will find 22 chapters, each of which provides a sure foothold for your climb to the top as a licensed master plumber. Chapter 1 will give you valuable advice on how to improve your odds for passing the licensing exam.

Chapters 2 through 22 are dedicated to specific points of the plumbing code. For example, chapter 7 concentrates on the code requirements for sanitary drainage systems. Chapter 12 deals with vents and venting. Each chapter concentrates on a specific section of the plumbing code. The chapters contain mock exams and tests that will reveal your knowledge of the code. These professionally designed exams and tests will give you insight into your ability to pass a real licensing exam.

As you take the tests in the following chapters, you will realize two goals. You will check your current knowledge of the code, and you will learn code requirements. The questions in the tests are written to simulate what you might expect to find on a real licensing exam. The tests are no stroll in the park, they're tough, and they're good. If you can pass these sample tests, you should pass the licensing exam.

How to use this book

I'd like to explain to you how to use this book effectively. Take the tests in each chapter and then check your answers with the correct answers at the end of each chapter. There is nothing formal or intimidating about it. You can test your knowledge privately and with complete control. There are, however, some aspects of the plumbing codes that I would like to discuss with you.

There is no single plumbing code that controls plumbers across the country. Instead, there are three major codes, and a few smaller codes. Basically, every state, county, and town can, to a degree, modify their plumbing code. Typically, one of the three major codes is adopted as the local code. Sometimes minor modifications are made to allow for local conditions. These modifications are generally minor, but they might confuse you, so let's talk a little more about how the various plumbing codes work.

This book is broken down by code zone. I've done this to make the book valuable for people who work under the jurisdiction of each of the three major codes. Refer to Tables 1-1 through 1-3 to determine your code zone.

The multiple-choice exams in the chapters are generic. By this, I mean that they are written to apply, for the most part, to all three major codes. The true-false tests in the chapters are written for specific code zones. For example, if you live in Virginia, you are in zone 3. You should take the true-false test for zone 3 and ignore the tests for zones 1 and 2. If you live in Florida, you should take the tests for zone 2 and ignore the tests written for zones 1 and 3. Take all of the multiple-choice exams, and take only the true-false test intended for your code zone.

When you have completed the tests, you will find the correct answers at the end of each chapter. Sometimes there will be more than one answer that might be considered correct. Under these conditions, choose the answer that most closely reflects the intent of the code. You will be prompted on this more in the individual chapters.

After you have checked your work, look up the questions that you answered incorrectly. The chapters of this book are broken down into the same basic categories used in the code books, and it will be easy for you to research your local code book for the correct answers.

Occasionally, you might have trouble finding information on a multiple-choice question in your code book. For example, the code in zone 1 does not dedicate a section of its code book to health-care plumbing. The code books in zones 2 and 3 do have specific sections that deal with health-care plumbing. If you live in zone 1 and take the multiple-choice exam on health-care plumbing, you might discover that your code doesn't contain the same rules and regulations as those covered in the exam. There are very few occasions when this sort of situation might arise, but there will be some.

I've tried to make this book as compatible with all three codes as possible, but as you might imagine, melding three codes together in one study guide is not an easy task. I heartily recommend that you obtain a copy of my other book, *National Plumbing Codes Handbook* (available from McGraw-Hill), to use as a companion guide to this book. In that book I discuss all three of the major codes and point out the differences. My code book is much easier to read and understand than most code books, and it is a valuable resource guide. Ideally, you should be working with a local code book, my code book, and this book to achieve maximum results from your study efforts.

I mentioned earlier that a small percentage of code jurisdictions use codes that are not entirely based on one of the three major codes. You can check with your local code enforcement office to confirm the code in use in your area. The chances are good, however, that you will be under the jurisdiction of one of the three major codes.

I've used the most current code rulings at the time of this writing. Some plumbing code aspects are revised periodically, and these revisions will occur after this book is published. The revisions are generally not of a major scale, so this book will be useful for years to come. It will be in your best interest, however, to study with a current local code book as you prepare for the licensing exam.

Errors and omissions are possible with any book. I've worked hard to ensure that the information in this book is accurate and dependable. However, I'm human, and you might find a mistake here and there. Please forgive me for any such errors or omissions. Neither this book or my code book is intended as a replacement for your local code book. My book is meant to help you interpret your local code; not to be your local code.

If you want to be a licensed master plumber, you can learn the code and pass the licensing exam. The work required to do this will be more exhaustive for some people than it is for others. Every individual works and learns at a different level. I'm sure, however, that with the help of this book, you will improve your odds for success in passing a licensing exam.

Now, let's get down to business, class is in session.

CHAPTER 1

A code-class instructor's recommendations

THIS book is comprised of a number of multiple-choice and true-false exams that will help you prepare to pass the licensing exam needed to obtain a master plumber's license. I have compiled these exams as a teacher of both apprenticeship and code-requirement classes at Central Maine Technical College in Auburn, Maine. I am a master plumber and have been in the profession for 20 years. As both a seasoned plumber and a teacher I have helped numerous plumbers pass their exams. I believe what I have learned will help you obtain your license as well.

Options

There are many options available to people who wish to become licensed plumbers. If you want to earn a plumber's license, you will have to learn the plumbing code. This can be done by attending several years of vocational school, or by working in the field and studying on your own. Both ways are effective, and different people might find one method better suits their personality.

When I started in the plumbing trade, I was a plumber's helper. A laborer would have been a better choice for my title, based on the type of work I did. My first six months of plumbing consisted of digging ditches, operating jackhammers, and carrying pipe, cast-iron pipe at that. During those six months, I lost 60 pounds, came home so tired I could hardly walk up the stairs, and nearly quit three times.

It was difficult to be impressed with the plumbing trade when it seemed all I did was grunt work.

When it was time for my six-month review, my supervisor and the company manager met with me. When they presented me with my performance report, I was impressed. It was then that I found out that the company tried to make life as hard as possible for rookies during their first six months of employment. The company wanted to make sure that the people they chose to train as plumbers had the grit and determination to stick with the program. I passed muster. From that day on, my plumbing career took turns for the better.

In the second half of my first year as a plumber's helper, I learned to solder joints, cut and thread pipe, pour lead joints, and in general, to install complete plumbing systems. My job sites ranged from remote residential jobs to a large health spa. The experience I gained gave me a sense of pride. By the end of my first year in the trade, I could plumb a new house without any help.

After I was taken out of the ditches and off jackhammer duty, I realized that I really enjoyed plumbing. The strange thing is, I still do, and it's been 20 years since I picked up my first plumbing shovel. Anyway, once I knew I wanted to be a plumber, I bought a code book. It was expensive, but I knew I would have to know and understand the code in order to be the best plumber in town.

My code book went almost everywhere with me. When I got to work early, I sat in my truck and studied the code. When I rode to the job sites with Jerry, the plumber who trained me, I asked him endless questions about the code. You see, then the language in the code book confused me and it helped a lot to have a knowledgeable plumber explain it to me.

I read my code book during my lunch breaks and on the way back to the shop, when I could get Jerry to drive. When I got home, I took a hot bath and read more of the code. You could probably build a case to show that I was obsessed with the plumbing code. I knew it was my ticket to making more money and having better working conditions. Of course, 20 years ago I had no idea how good the plumbing trade was going to be to me.

What does a plumber do who has put in 20 years of field service? Well, I write books, I build houses now and then, I dabble in real estate, I offer training and development courses, and, of course, I still run a plumbing business. Do I still do plumbing myself? Yes, I do. Do I have to go out and do plumbing with my own hands? No, I could make enough money to be comfortable without touching a piece of pipe, but I enjoy plumbing, so I still get out there and do it.

Not everyone who comes into the trade enjoys the work as much as I do, but most of the plumbers I've known have been satisfied with their career choice. One of the questions I get asked most often is that of how much money does a plumber really make. This is a difficult question to answer, since not all plumbers make the same amount of money. I can say this. If you are a good, experienced plumber, you can make a better-than-average living, and there is almost always someplace where you can work, even in the worst economic times.

I would like to tell you more about what you might expect from your career as a plumber, but there isn't enough space in this book to allow me that freedom. However, I have written another book, also published by McGraw-Hill, that might be of interest to you. *The Plumbing Apprentice Handbook* is an in-depth look at exactly what you are getting into when you opt to become a plumber. While we are on the subject of other books, I also wrote the *National Plumbing Codes Handbook*, which makes the plumbing code easier to understand. Owning the *Codes Handbook* is kind of like having an experienced plumber with you to answer all of your code questions. It, too, is published by McGraw-Hill. Now, let's look at all of your exam preparation options, on a one-by-one basis.

The hard way

Trying to study a code book on your own is the hard way to learn the plumbing code. The plumbing code can be very difficult to understand, unless someone explains it and guides you through it.

However you choose to prepare for the licensing exam, you will need a local code book. Make sure the book you buy is current. Plumbing codes change, and an old book might result in a failed exam. Once you've got your code book, you're probably going to need some help understanding it. My book, *National Plumbing Codes Handbook*, can help get you through the tough spots of the code.

Self-study

Self-study is the least expensive way to prepare for the licensing exam. Code-preparation classes, while not extremely expensive, cost money. With a local code book and this book, you can teach yourself the plumbing code. Your code book will provide the facts and foundation of the plumbing code, and this book will test your knowledge and help alert you to areas on which you require further study.

Code classes

Code classes exist to help plumbers and aspiring plumbers learn and understand the code. If the class instructor is good, the price of admission is money well spent. There is no substitute for being able to question an experienced instructor about confusing areas of the plumbing code. When I was learning the trade, code classes didn't exist, at least not in my area or to my knowledge. If code-preparation classes had been available to me, I would have attended.

I don't know how many people pass the licensing exams on their first attempt. Experience has shown me that a lot of plumbers and would-be plumbers fail the exam on their first attempt. I passed all of my tests on the first try, and the students who have attended my code classes have done very well in passing their tests on the first run.

Code classes are not a prerequisite for exam preparation, but I feel they are a good value and an investment in your professional future. If you combine a good code class with extensive self-study, you can hardly lose.

The buddy system

If you can't afford a code class, perhaps you can study with the buddy system. If you know someone else who is preparing for the licensing exam, the two of you can work together. Take turns being instructor and student. Test each other's knowledge with the exams in this book. Working together can achieve your goal quicker than working alone, and the competitiveness between you and your partner might make you put a little extra effort into your studies.

Learn it and understand it

The key to passing a licensing exam is the plumbing code. You must learn it and understand it. Learning the code is one thing, understanding it can be quite another. The way that licensing exams are worded can be quite tricky, even for experienced plumbers. If you think you can memorize the code book and pass your test, you might be sadly mistaken. Don't stop studying the code until you understand it. Let me give you a few examples of what I'm talking about here.

The plumbing code is filled with exceptions and key words. The key words often give people the most trouble. For instance, let's assume that the code says, "All toilets shall have discharge openings that are no less than 2 inches in diameter."

Now, suppose you saw this question on a true-false test: "The discharge opening of a toilet should be 2 inches in diameter." Would the answer to this test question be true or false? It would be false. The code implies that the opening must be no smaller than 2 inches. The test question suggests that a 2-inch opening is preferable, but not mandatory. Simple little plays on words, such as I've just shown you, can ruin your test score.

The exception clauses in the code can provide as much trouble for you as the mixture of key words. For example, assume a passage in the code reads like this, "All local vents serving bedpan washers, except where only one bedpan washer is being served, must be trapped." This is a simple little passage, but watch what it can do to you on a true-false test. Here's the test question: "All local vents for bedpan washers must be trapped." Well, you know the answer is false, because we have just discussed it. But if you were sitting for the licensing exam, you might give an incorrect answer because all vents for bedpan washers must be trapped, unless the vent is serving a single fixture. See how tricky things can get?

Once you have learned the code and understand what you've learned, applying it to test questions or work in the field will be much easier. I can't stress enough that you must do more than memorize the words in the code book. Take the time to learn and understand the code before you sit for the exam.

Develop a study plan

Develop a study plan that allows you to prepare properly for the licensing exam without interfering too much with your daily routines. If you are working for a plumbing company, you have a big advantage over someone who is working outside of the trade. Working with plumbing on a daily basis allows you to absorb a certain amount of knowledge. This, however, can be more of a hindrance than a help. What do I mean by this? I'm saying that a person who works in the trade might take things for granted or be exposed to bad plumbing practices.

People studying for a master plumber's license are not as likely to get in trouble from their bad habits in the field as journeyman plumbers. Even so, if you're working with a plumber in the field, you might learn techniques that will influence the outcome of the licensing exam. This is not a perfect world in which we work, and not all plumbers work within the parameters of the plumbing code. If you learn from a plumber who cheats the system, you might come up short on the licensing exam.

Journeyman plumbers who want to progress to the master level must often block out their day-to-day plumbing procedures in order to pass the licensing exam. Work in the field is not always done by the book, but to pass a licensing exam, the answers must come from the book.

Now, I'm not talking about blatant abuses of the code, but of little discrepancies that might result in wrong answers. For example, the plumbing inspector in your area might not require a 24-hour notice to schedule an inspection. This is fine on a daily basis, but if you work in zone 1 or zone 3, you'd better remember that a 24-hour notice is required to schedule an inspection.

The point to all of this is that you must shut out what you see in the field and concentrate on what the plumbing code requires. The fact is, to pass the test, you must go by the book. What happens after you are licensed is between you, your employer, and the local code enforcement office. But, for now, you must go by the book.

In order to go by the book, you must understand the book. Set aside some specific study time. It can be first thing in the morning, your lunch break, after supper, or just before bed. When you study doesn't matter, but it does matter that you do study. If you don't set up specific times and places to study, you are less likely to stick with the program, which might result in you failing the licensing exam. Determine how much time you can devote each day to studying, and make a study schedule that you can use as motivation.

Create a comfortable atmosphere

To make the most of your study time create a comfortable atmosphere in which to study. If you try to absorb a complicated section of the code while children play loudly around you, there is a good chance you will not capture the true meaning of the code. Reading a code book can be difficult enough under ideal conditions, but when you put yourself in a noisy environment your chances for success decrease.

Where should you study? The location you choose as a study area should be one in which you are in control. The exact location might be an office, a bedroom, a den, a public library, your truck, or even a travel trailer, if you happen to have one. Find a place that is quiet and that will provide you with undisturbed time for studying.

My office is quiet during the day, except for the ringing of the telephone. However, at night, when my wife and daughter get home, my office is not as conducive to productive work. If I'm working on a project that requires a lot of concentration, I leave my office. My retreats have included a tent, a travel trailer, a comfortable spot under a pine tree, and remote wilderness cabins. These secluded settings have afforded me the privacy I needed to accomplish complicated tasks. I'm sure you can find some location that will do the same for you.

Get in the proper mind-set

People who work all day and have family responsibilities might find it troublesome to set aside their other responsibilities to study the plumbing code at night and on weekends. If you have this problem, you're going to have to overcome it.

There are many different ways to gear your mind to studying. Some people like to exercise just prior to sitting down to a study session. Others prefer to meditate for awhile, and still others seem to do their best studying while listening to a radio.

I've always found myself to be at my best late at night. Many people get tired in the evening and don't have the attention span needed to study or complete complex jobs. I, however, seem to work best during the late night and early morning hours. The phone doesn't ring, my family is asleep, and I can concentrate all of my efforts on whatever it is that I'm doing. I can't tell you what method will work best for you, but it is necessary for you to clear out the daily tribulations so that your mind can retain the plumbing code.

Don't wait

Don't wait until the last few days to start studying. The amount of information you need to learn is significant. If you wait until the last minute to begin studying, you are setting yourself up for failure.

I've stressed that learning the code is not enough, that you must also understand it. I know some people, like my wife, who prefer to cram for a test just before the test date. They memorize the material for the test, and then much of it is forgotten in the weeks to follow. I don't study this way, and I don't believe the memorization method will work consistently when you take licensing exams. Due to the way the test questions are worded, it is necessary to understand fully the question and the intent of the code in order to provide a correct answer.

My feelings on the memorization method are based on personal experience, both as an individual taking licensing exams and as an instructor of exam-preparation courses. I've conducted experiments in my code classes that have proven that tricky wording often catches students who cram for their exam. These same students seem to do fairly well when the test questions mirror the code, but when the questions are disguised, as they often are on licensing exams, the students tend to receive much lower scores.

Don't fall into a routine

If you go through your code book the same way, time and time again, you will become conditioned to giving specific answers almost out of habit or reflex. This reflex motion can destroy your score on the licensing exam.

I have been a victim of routine while studying the code. For example, I used to know the sizing tables in the code by heart, as long as I could quote them in the order in which I learned them. I learned the tables from the smallest pipe sizes to the largest. This allowed me to spout off the proper vent-to-trap distances and sizing requirements with ease. There was, however, a problem with this procedure.

After I had studied my code book until I thought I knew it from cover to cover, I asked a friend of mine to quiz me on the contents. As I sat poised to provide quick answers to any questions I was asked, I suddenly realized I didn't really know the code. Oh, I knew it by heart, but only in my own way. When my friend started asking questions that were out of sequence with the way that I had studied the book, I made a lot of mistakes. Fortunately, I found my fallacies before the day of the licensing exam and was able to correct them.

Whether you are studying definitions, tables, charts, or piping materials, don't zoom in on one topic for too long. Move around the chapters in your code book, and learn them a little here and a little there. Split up your study procedure and you will gain two benefits. You will not get as bored with your studies, and you will not be as likely to fall into a detrimental pattern the day of your exam.

Don't panic

A lot of people know the material, but their minds seem to go blank when they must take a test. I've never had this problem personally, but I've known a number of people who suffer from pre-test trauma. If you allow yourself to get nervous or upset just before you sit for the exam, you are dooming yourself to failure.

How can you avoid a panic attack? The best way I know to control pre-test jitters is to be well prepared for the test. If you know the material inside and out, you are going to pass the exam. You must, however, have confidence in your knowledge. If there is any doubt in your mind about how proficient you are with the plumbing code, there will be room for fear. Eliminate the fear with proper preparation.

Taking the tests in this book is one excellent way to overcome the fear of failing the licensing exam. The questions in the chapters that follow are more difficult in many ways than those on the actual exams. The tests in this book ask questions that you are not likely to see on the real test, at least not in the quantity I have provided in the various chapters. If you can pass the tests in this book, you should be more than capable of passing the licensing exam.

Build your confidence by taking the tests offered in this book and by having a friend test you orally. Get your friend to thumb through your code book and ask random questions. If you can stand up to random testing and the tests in this book, you should have the confidence to beat the fear factor on the day of the real test.

Create your own tests

When you create your own tests, you learn the code almost by accident. In order to craft good questions, you are forced to read and understand the language in your code book. This is exactly what you should be doing as a study plan, but it comes easier when you don't realize you are studying. Drafting your own mock tests will not seem like formal studying, but it can produce some of the most effective results.

After you have put together a suitable number of test questions, let them sit for several days. It is a good idea to record the proper answers on a separate piece of paper for reference after you have taken your own test. When the tests have cured for a few days, sit down and try to answer the questions correctly. Put your answers on a separate piece of paper. This will allow you to take the same tests over again as the real test date draws nearer. Since you created the test, there should be very little intimidation involved with the process.

Once you have taken your own tests and scored them, you will probably be pleased with your new knowledge. Sample tests, like the ones in this book, are one of the most effective methods you can employ to learn the code. By combining self-made tests with the tests I'm providing in this book, you should obtain excellent coverage of your local code, and your knowledge should grow quickly.

Step away from it

If you study the code too intensely, you will suffer from burnout. Sometimes it is best to step away from it for awhile and then go back to studying with a fresh attitude. If you force yourself to study for two hours every night, you might find that you resent the code and the demand on your time. If you are not receptive to learning, you will not learn. This problem can be avoided by giving yourself enough time to study, without having to study in a gruelling manner.

Don't get lazy

While it is fine, and even beneficial, to step away from your studies from time to time, discipline yourself to some study structure. It is easy to put off your studies until the next day, but there are only so many times that you can do that without coming up short. If you get lazy, you are either going to panic and cram your studies in at the last minute, or you are going to have a tough time passing the exam.

Check for local variations

Before you become too comfortable with your knowledge of the code, check for local variations. Regardless of where you live or work, the plumbing code for that area might be an amended version of the primary code. Each code jurisdiction has the right to amend and interpret the plumbing code. These amendments and interpretations can contradict the primary code. While most areas don't make major changes to the plumbing code, it is well worth your time to make sure that you are studying material that will be on the licensing exam. This is especially true for individuals who will be taking exams in areas where two different codes meet or overlap.

Ask questions

One of the best ways to learn about something new is to ask questions. This fact applies to learning the plumbing code. The code can be difficult to decipher. If there are parts of the code that you don't understand, consult my book, *National Plumbing Codes Handbook*, or ask a master plumber or plumbing inspector to verify your interpretation of the code. If you are already working for a plumbing company, it will be easy to ask your master plumber questions that might arise as you study for the licensing exam. If you don't have immediate access to a master plumber, record the questions you have about the code. When you have a list together, talk with your local plumbing inspector.

While we are on the subject of asking questions, ask some questions about how the licensing exam will be given. Will you be allowed to use a portable calculator? Most test sites will let you use a calculator, as long as it is not equipped with paper for printouts. Should you bring pencils to the test? Normally, testing requires the use of Number-2 pencils, which the individuals being tested are required to bring. Should you bring scratch paper? Most test facilities won't allow you to bring paper into the test center. The testing facilitators, however, will often provide paper for those who request it. When will you know the test results? Many test sites can give you an unofficial test score the day you take the exam, but the official scores are typically mailed to you at a later date. These are just some of the questions you might want to ask before you sit for the exam. The more you know about what to expect, the less you will have to fear.

What's next?

The chapters that follow include two types of tests. The first test in each chapter is a multiple-choice exam. The length of these exams varies, depending on the subject matter covered. Most multiple-choice exams are followed by true-false tests. The multiple-choice exams are somewhat generic, while the true-false tests are geared to specific codes for various areas. Complete both the multiple-choice exam and the true-false test for your region before you check your responses against the correct answers located at the end of each chapter.

Keep in mind that since the three major plumbing codes are not identical, there might be some questions in the generic multiple-choice exams that you might have difficulty locating in your code book. For example, the code serving zone 1 does not discuss health care plumbing specifically. The codes in zones 2 and 3 do have specific sections on this subject. Since the code book for zone 1 doesn't break out this information, students in that zone will have some difficulty with the test questions on health care plumbing. In most cases, however, you will not have any trouble comparing the test questions in this book with the various sections of your code book. Now, let the tests begin (Table 1-1, Table 1-2, Table 1-3).

Table 1-1

States in Zone 1

Washington	Iowa
Oregon	Nebraska
California	Kansas
Nevada	Utah
Idaho	Arizona
Montana	Colorado
Wyoming	New Mexico
North Dakota	Indiana
South Dakota	Parts of Texas
Minnesota	

Table 1-2

States in Zone 2

Alabama	Florida
Arkansas	South Carolina
Louisiana	Parts of Texas
Tennessee	Parts of Maryland
North Carolina	Parts of Delaware
Mississippi	Parts of Oklahoma
Georgia	Parts of West Virginia

Table 1-3

States in Zone 3

Virginia	Vermont
Kentucky	New Hampshire
Missouri	Rhode Island
Illinois	New Jersey
Michigan	Parts of Delaware
Ohio	Parts of West Virginia
Pennsylvania	Parts of Maine
New York	Parts of Maryland
Connecticut	Parts of Oklahoma
Massachusetts	

CHAPTER 2

Administrative and enforcement issues

ADMINISTRATIVE and enforcement issues pertain to both journeyman and master plumbers. The rules for master plumbers are a bit more complex and comprehensive than they are for journeyman plumbers. This does not mean that journeyman plumbers have no responsibilities under the rules that pertain to administration and enforcement issues. In fact, this is the section of the code book that relates to a person's need to be properly licensed in order to perform plumbing work.

When you sit for the exam, you are going to be allowed to answer a certain percentage of the questions incorrectly. If you exceed the percentage of incorrect answers allowed, you will not pass the test. A question about administration is just as important as a question about chemical wastes when it comes time to tally up the right and wrong answers. A miss is a miss, and a failing score is the same no matter why you failed. I tell you this in hopes that you won't underestimate the administration and enforcement section of your exam.

When you open a plumbing code book, the first part of the book deals with administration. Why is this? Because it is the administration of the code that enforces the rest of the code requirements. To work within the parameters of a plumbing code, you must understand how the code is administered and enforced. This chapter is going to test your knowledge of the areas of administration and enforcement.

Will there be questions about administration and enforcement on the license exam? It's a safe bet to say yes. Most plumbing exams are altered to some extent on a rotating basis to keep people from cheating, but issues of administration and enforcement are common topics found on all plumbing exams. There is no way of knowing how many of the questions on the exam will come from the administration section of the code. Each plumbing exam will be different. The only way to be properly prepared for such an exam is to know all the material that might be presented on the test.

As a code instructor, I have taught a lot of people how to interpret the plumbing code and how to prepare for a licensing exam. Some of the people have been apprentices preparing for the journeyman license, others have been journeyman plumbers preparing for the exam to obtain a master's license. It is amazing how few licensed plumbers have a good working knowledge of administration and enforcement issues.

I've trained a lot of plumbers, both as a master plumber and as a code instructor, and I'm sure that a majority of them had little respect for the administration section of the code book. It's not that they didn't intend to comply with the plumbing code, it is just that they didn't seem to believe the code's administration aspect was important enough to learn for the exams. The section is pertinent to all licensed plumbers, especially master plumbers who are the ones who usually deal with administration issues.

Before you answer the sample test questions, remember that there are three major plumbing codes. Each code offers its own code book and code requirements. Many of the rules and regulations are identical, but some of them are different. This book is intended to prepare you for an exam created from any one of these three codes. While it is possible to create sample tests that indicate your knowledge of one code, it is not possible to plan for local code modifications.

Every local jurisdiction is allowed to adapt and alter the plumbing code. Changes made on the local level don't show up in a typical code book. Before you become completely comfortable with your grasp of the code, check with your local code enforcement office to see if there are any special local rules or regulations that might be in conflict with what the code book states. Local changes are usually minor and they normally will not be extensive enough to impact your performance on the exam. But, you will be wise to confirm this fact before you take the test. Now, let's get on with the sample questions.

Multiple-choice exam

1) When a matter that is essential for sanitary safety pertaining to an existing building is not covered in the code book, who has the authority to make a decision on how the matter should be handled?

A) A master plumber
B) A journeyman plumber
C) A code official
D) None of the above

2) When a matter that is essential for sanitary safety pertaining to a new building is not covered in the code book, who has the authority to make a decision on how the matter should be handled?

A) A master plumber
B) A journeyman plumber
C) A code official
D) None of the above

3) If a part of a building is in violation of the code and a code officer condemns the building, can the building continue to be used without violating the code?

A) Yes
B) No
C) Yes, so long as there is a written agreement
D) Yes, so long as the owner of the property agrees to correct the code violation within 14 days

4) Which of the following aspects of plumbing systems are governed by the plumbing code?

A) Sanitary drainage
B) Storm drainage
C) Sanitary facilities
D) All of the above

5) Which of the following aspects of plumbing systems are governed by the plumbing code?

A) Water supplies
B) Storm water
C) Sewage disposal
D) All of the above

6) Which of the following is the intent of the plumbing code?

A) To produce revenue for the local jurisdiction
B) To ensure public safety, health, and welfare
C) To regulate plumbers
D) All of the above

7) When is an alteration or repair to an existing plumbing system in violation of the plumbing code?

A) When it is done without a permit
B) When it is not done by a master plumber
C) When it renders the plumbing system unsafe
D) Any of the above

8) When is an alteration or repair to an existing plumbing system in violation of the plumbing code?

A) When it is done without prior code approval
B) When it is not done by a master plumber
C) When it adversely affects the performance of the plumbing system
D) Any of the above

9) This question calls for you to assume a certain scenario and make a decision. Here is the situation: It is your job to add a bathroom in a home where two full bathrooms presently exist. The building drain and sewer for them are both 3-inch pipe. Current code regulations limit the installation of water closets on a 3-inch pipe to no more than 2 toilets. However, you are working with existing plumbing, so do you have to change the size of the existing pipes or can you connect to them with the new toilet? The plumbing code allows some exceptions from current rules for existing plumbing that is being altered. If you found yourself in this position, what would you do? Your decision and the ruling in the code book might not be identical, so answer this question with the correct answer as it appears in the code.

A) Consult the local code officer for a ruling
B) Connect to the existing 3-inch pipe with the new toilet, since the existing pipe is grandfathered in the code
C) Increase the size of the building drain or sewer from the point of connection that exceeds the limit of two water closets
D) Ask your boss what to do

10) This question also puts you in a specific situation. The owner of a building that was used previously as a retail store has requested that the plumbing in the building be converted to accommodate a hair salon. What are you required, by code, to do? Answer this question with the best possible answer. More than one of the answers might apply, but only one is the best answer.

A) Gain approval and certification for the proposed changes
B) Install proper back flow prevention devices
C) Defer to a master plumber
D) Use materials approved by the local code enforcement office

11) All plumbing repairs and replacements are required to be made in what manner?

A) By a master plumber
B) In compliance with the local plumbing code
C) By a journeyman plumber
D) Both A and B

12) All plumbing repairs and replacements are required to be made in what manner?

A) By a master plumber
B) In compliance with the local plumbing code
C) In a safe and sanitary manner
D) Both B and C

13) When minor repairs or replacements are made to an existing plumbing system, which of the following rules apply?

A) All work must be brought up to current code requirements for new plumbing
B) The repair or replacement may be made in the same manner and arrangement as in the existing system
C) A permit must be obtained
D) Both B and C

14) All plumbing systems, both new and existing, must be maintained to assure which of the following:

A) That the system is upgraded periodically to comply with changes in the plumbing code
B) That the system remains in good working order and continues to be both safe and sanitary

C) That no unlicensed individuals alter the system in any way
D) All of the above

15) Who is responsible for the maintenance of a plumbing system?
A) A master plumber
B) The property owner
C) The property owner's designated agent
D) Either B or C

16) If a building is scheduled to be demolished, who is responsible for making proper notifications to the people or agencies that deserve to be notified?
A) A master plumber
B) The property owner
C) The property owner's designated agent
D) Either B or C

17) If a building is scheduled to be demolished, who is to be notified of the intended work?
A) A master plumber
B) The property owner
C) The property owner's designated agent
D) The principals of utilities having service connections involving the building

18) Who has a right to modify the provisions of the local plumbing code?
A) The mayor
B) A master plumber
C) A code official
D) Any of the above

19) Variances to the code are sometimes approved if it is felt by the proper authorities that there is just cause for the request. If a variance is requested, who is required to make the request?
A) A master plumber
B) A property owner
C) A property owner's representative
D) An architect
E) Either A or D
F) Either B or C

20) When a variance to a code requirement is requested, how should the request be made?

A) In person
B) By phone
C) In writing
D) In the presence of the appeal board

21) When a variance from the plumbing code is requested, it might be approved if the alteration is considered to be in the spirit and intent of the code and does not represent a risk to public health, safety, and welfare. Who has the authority to grant a variance?

A) The clerk of the court
B) The mayor
C) A code official
D) An independent engineering firm

22) Who has the authority to authorize the use of previously used plumbing materials and equipment?

A) A master plumber
B) A property owner
C) A property owner's representative
D) A local code official

23) Before used plumbing materials or equipment can be approved for use in new applications, the materials or equipment must be:

A) Reconditioned
B) Tested
C) Placed in working order
D) All of the above

24) Who has the authority to allow the use of materials and equipment that are not covered in the code book?

A) A property owner
B) A property owner's representative
C) An independent engineering firm
D) A code official

25) For an alternative material or piece of equipment to be approved for use, it must meet which of the following criteria:

A) The design must be satisfactory
B) It must comply with the intent of the code

C) It must be approved by the property owner
D) Both A and B
E) Both B and C
F) It must be approved by the Master Plumber's Guild

26) For an alternative type of material or equipment to be approved for use, it must be considered along which of the following guidelines:
A) Quality
B) Strength
C) Effectiveness
D) Fire resistance
E) Durability
F) Safety
G) All of the above
H) Both A and E

27) If an alternative material is submitted for approval, the local code official might request that research and investigation be conducted to prove the material suitable for use. Who is responsible for the cost of such research and investigation?
A) The master plumber
B) The property owner
C) The applicant of the request for acceptance
D) The code enforcement office

28) According to the plumbing code, who is the executive official in charge of the code?
A) The code official
B) The clerk of the court
C) The mayor
D) The town council

29) Who is responsible for code-compliance inspections?
A) The master plumber
B) The property owner
C) The code official
D) The fire marshall

30) Who is responsible for keeping records of applications received, permits and certificates issued, fees collected, reports of inspections made, and notices and orders issued in regard to the plumbing code?
A) The health department
B) The public works department
C) The local code official
D) The master plumber involved in the work being performed

31) When a plumbing permit is required for proposed work, the work may not be commenced until:

A) A permit is applied for
B) A permit is issued
C) A permit is approved
D) A permit is signed by a code official

32) Which of the following types of work requires a plumbing permit?

A) The replacement of an electric water heater
B) The repair of a pipe that froze and split
C) The replacement of a faucet stem
D) The clearance of a drain stoppage
E) The replacement of a faulty gate valve

33) Which of the following individuals will not normally be eligible to receive a plumbing permit?

A) A master plumber
B) A journeyman plumber
C) A property owner who will personally perform all plumbing work described in the permit application on a primary residence owned by that person
D) Both B and C

34) A plumbing permit is not valid until it is:

A) Posted in a conspicuous place on the job site
B) Signed by the master plumber, or an authorized representative
C) Signed by the code official, or an authorized representative
D) Both B and C

35) A journeyman plumber who is installing plumbing under a permit obtained by a master plumber must adhere to which of the following rules:

A) All work must be done under the direct supervision of a master plumber
B) All work must be done in compliance with the plumbing code
C) All work must be done in the manner as it was expressed to the code official at the time of application for the permit

D) All of the above
E) Both A and B
F) Both B and C

36) When a stop-work order is issued by a code officer, you must immediately do what?
A) Inform the master plumber
B) Quit working on the job
C) Request a written explanation for the order
D) Finish out the day and deliver the notice to your master plumber

37) Plumbers are required to have what records accessible to the local code official at all times?
A) Material invoices
B) Test results from previous inspections
C) Previous inspection results
D) Both B and C

38) Plumbers are required to perform all plumbing work in which of the following ways:
A) A workmanlike manner
B) An acceptable manner
C) Under the direction of the property owner
D) All of the above
E) Both A and B
F) Both A and C

39) When will a code officer issue a notice of approval for a plumbing system?
A) When all work is complete
B) When all tests and inspections prove the work is in compliance with the local plumbing code
C) When requested, in writing, to do so
D) When the owner of a property requests a notice of approval

40) According to the plumbing code, what constitutes good workmanship?
A) Any work done by a licensed plumber
B) Any work that is clean and neat

C) Any work that secures the results intended by the plumbing code
D) Any work acceptable to the property owner

41) Work that requires a plumbing permit may not be started until _____. Fill in the blank.

A) A permit is applied for
B) A permit is issued
C) Either A or B
D) Neither A nor B

42) A permit is not required for repairs that involve only _____. Fill in the blank.

A) The relocation of fixtures
B) The working parts of faucets and valves
C) The replacement of water heaters
D) None of the above

43) A permit is not required for repairs that involve only _____. Fill in the blank.

A) The clearing of stoppages
B) The working parts of faucets and valves
C) Both A and B
D) None of the above

44) A permit is not required for repairs that involve only _____. Fill in the blank.

A) The clearing of stoppages
B) The working parts of faucets and valves
C) The replacement of defective valves or faucets
D) All of the above

45) Which of the following persons are allowed to apply for a plumbing permit?

A) Journeyman plumbers
B) Master plumbers
C) The general public
D) None of the above

46) Which of the following persons are allowed to apply for a plumbing permit?

A) Registered design professionals
B) Master plumbers

C) Property owners
D) Any of the above

47) Which of the following persons are allowed to apply for a plumbing permit?

A) Lessees of buildings
B) Master plumbers
C) Agents of property owners
D) Any of the above

48) Which of the following are key elements that must be included in an application for a plumbing permit?

A) The number of fixtures to be installed
B) The location of the work
C) The occupancy of the building in which the work is to be done
D) All of the above

49) How many sets of construction documents must be provided during the application for a plumbing permit?

A) 1
B) 2
C) 3
D) 4

50) Which of the following elements is a substantial part of the construction documents required in order to obtain a plumbing permit?

A) Floor plans
B) Riser diagrams
C) Both A and B
D) Neither A nor B

51) Construction documents presented with a permit application might be required to provide information on _____. Fill in the blank.

A) The direction of flow
B) The quantity of flow at a specific time of the day
C) The individual plumber who will install the plumbing system
D) None of the above

52) Construction documents presented with a permit application might be required to provide information on _____. Fill in the blank.

A) The grade of horizontal piping
B) The elevations relevant to the proposed plumbing
C) Both A and B
D) None of the above

53) Construction documents presented with a permit application might be required to provide information on _____. Fill in the blank.

A) The drainage fixture-unit load on the system
B) The supply fixture-unit load on the system
C) Both A and B
D) None of the above

54) A site plan will normally show the location of _____. Fill in the blank.

A) The proposed water service
B) The proposed sewer connection
C) Both A and B
D) None of the above

55) Vent stack terminations should be shown on a site plan with respect to _____. Fill in the blank.

A) The number of plumbing fixtures installed
B) The type of plumbing fixtures installed
C) Building ventilation openings
D) None of the above

56) A permit application will be considered abandoned if the application is not diligently prosecuted or a permit is issued within a _____ month period of time. Fill in the blank.

A) 3
B) 6
C) 9
D) 12

57) A code official can extend the time allowed prior to a permit application becoming abandoned by a period of _____ days. Fill in the blank.

A) 30
B) 60
C) 90
D) 120

58) Any plumbing permit issued might become invalid if work is not started within _____ months from the date the permit was issued. Fill in the blank.

A) 3
B) 6
C) 9
D) 12

59) Any plumbing permit issued might become invalid if work ceases for a period of time that exceeds _____ months. Fill in the blank.

A) 3
B) 6
C) 9
D) 12

60) All approved construction documents will be _____. Fill in the blank.

A) Stamped by a code official
B) Endorsed by a code official
C) Both A and B
D) Either A or B

61) Which of the following is likely to be located on a site plan?

A) A riser diagram
B) A stack diagram
C) The location of a private sewage disposal system, if a public sewer is not available
D) All of the above

62) Which of the following locations are required to be supplied with a set of approved construction documents?

A) The code enforcement office
B) The job site
C) Both A and B
D) Either A or B

63) If a false statement is made during the application for a permit, what is likely to happen?

A) The permit will be suspended
B) The permit will be revoked
C) The master plumber will go to jail
D) The property owner will be sued

64) Prior to construction documents being approved by a code official, which of the following might be allowed?

A) All plumbing work can be installed, so long as a permit is applied for within 90 days of the completion of work
B) A code officer can authorize a partial approval, allowing some of the work to be done

C) A master plumber must apply to a board of appeals
D) None of the above

65) How much notice is required to be given to a code official before work under a permit is started?

A) 8 hours
B) 24 hours
C) 48 hours
D) 72 hours

66) Where must a true copy of a permit be kept during the entire time work is being done under that permit?

A) Displayed in the master plumber's office
B) At the code enforcement office
C) On the job site
D) Any of the above

67) Under what conditions does a code officer have the authority to order and require occupants of a building to vacate a structure forthwith?

A) Code officers do not possess such a power
B) When a senior code officer approves the action
C) When in the opinion of the code officer a situation exists where imminent danger of contamination or a sanitation hazard might endanger life
D) When a code violation is discovered

68) Of the following, who can request an appeal from a board of appeals?

A) A permit holder
B) A property owner
C) A master plumber
D) Any of the above

69) Which of the following is a suitable cause for requesting an appeal?

A) A dislike for the present plumbing code
B) A denied permit application
C) A claim that the true intent of the code has been interpreted incorrectly
D) Any of the above

70) Which of the following is a suitable cause for requesting an appeal?

A) An opinion that the provisions of the code do not apply fully
B) A denied permit application
C) A claim that the true intent of the code has been interpreted incorrectly
D) Both A and C

71) Which of the following is a suitable cause for requesting an appeal?

A) An opinion that the provisions of the code do not apply fully
B) When an equally good or better form of construction is proposed
C) A claim that the true intent of the code has been interpreted incorrectly
D) Any of the above

72) Where are forms for an appeal obtained?

A) From the appeals board
B) From a code officer
C) From a post office
D) Any of the above

73) Within how many days are you allowed to file for an appeal after notice has been served?

A) 3
B) 5
C) 10
D) 20

74) How many people are required to form a board of appeals?

A) 3
B) 5
C) 7
D) 9

75) Under what conditions is a member of the board of appeals unable to hear a case?

A) When the member has a personal interest in the case
B) When the member has a professional interest in the case
C) When the member has a financial interest in the case
D) Any of the above

We have reached the end of the multiple-choice exam. Don't turn to the end of the chapter to check the results of your answers at this time. Instead, proceed directly to the true-false questions that pertain to your region and answer those questions. When you have completed all the test questions that pertain to your region, turn to the answers and check your work.

True-false test for zone 1

1) The plumbing code is in existence to provide a minimum requirement of standards for the protection of public health, safety, and welfare.

   True False

2) The plumbing code applies to the erection, installation, alteration, repair, relocation, replacement, addition to, use or maintenance of plumbing systems.

   True False

3) Any alteration to an existing plumbing system requires that the present system be upgraded to current code requirements.

   True False

4) The academic authority shall have the right to duly appoint and enforce this plumbing code.

   True False

5) A plumbing permit is not required to fix leaks in drains that are not concealed.

   True False

6) If a concealed trap or drain becomes defective and must be replaced, the job shall be considered as new work and a permit will be required.

   True False

7) The removal and reinstallation of a water closet does not require a permit, so long as the work does not require the replacement or alteration of valves and pipes.

   True False

8) A set of approved plans and specifications, which have been accepted by the code official, must be kept in the office of the permit holder at all times.

True False

9) The administrative authority may require a notice of at least one work day prior to making an on-site plumbing inspection.

True False

10) It is a violation of the plumbing code to connect to a water supply line or sewer without the permission of the administrative authority.

True False

True-false test for zone 2

1) If work is started prior to the issuance of a plumbing permit, a penalty equal to 100 percent of the normal permit fee may be assessed when the proper permit is acquired.

True False

2) Plumbing code officials are not required to make all inspections of plumbing systems personally, they may rely on the reports of approved inspection agencies.

True False

3) All plumbing permits must be posted in a conspicuous place on the site of the job prior to starting work.

True False

4) Posted plumbing permits must remain in a conspicuous place only until approval has been granted for rough-in plumbing.

True False

5) The trenches for underground plumbing may be backfilled prior to a site inspection by a code official.

True False

6) Rough-in plumbing must be inspected and approved by a code official before it may be concealed.

True False

7) All testing of plumbing systems shall be done at the expense of the owner or permit holder.

True False

8) A code official has the authority to authorize a temporary connection to a utility, such as a water main, for the purpose of testing a plumbing system.

True False

9) The final inspection of a plumbing system will normally be done when all plumbing is installed and operational.

True False

10) A plumbing permit is construed to be a license to proceed with the work authorized by a code official.

True False

True-false test for zone 3

1) Buildings being moved into or within the jurisdiction of the plumbing code are exempt from current code regulations.

True False

2) Used materials are not allowed in the installation, repair, or alteration of plumbing systems.

True False

3) A plumbing permit must be obtained before a faucet can be replaced. A faucet can be repaired without a permit, but it may not be replaced until a permit is obtained.

True False

4) Every permit shall be signed by an authorized code official.

True False

5) Any permit issued will become invalid if the authorized work is not started within 4 months from the time a permit was issued.

True False

6) Code officials have the right to enter any building within the local jurisdiction in the course of discharging official duties at any reasonable hour of the day.

True False

7) Alterations to existing plumbing systems must be made in a safe and sanitary way.

True False

8) Code officers have the authority to interpret the plumbing code and make modifications as they deem necessary.

True False

9) A release must be signed by all utility companies having connections made to or within a building prior to its demolition.

True False

10) Journeyman plumbers are responsible for the safe and sanitary maintenance of a plumbing system at all times.

True False

Answers to multiple-choice exam

1) C	8) C	15) D	22) D
2) C	9) C	16) D	23) D
3) B	10) A	17) D	24) D
4) D	11) B	18) C	25) D
5) D	12) D	19) F	26) G
6) B	13) B	20) C	27) C
7) C	14) B	21) C	28) A

29) C	41) B	53) C	65) B
30) C	42) B	54) C	66) C
31) B	43) C	55) C	67) C
32) A	44) D	56) B	68) D
33) B	45) B	57) C	69) C
34) C	46) D	58) B	70) D
35) F	47) D	59) B	71) D
36) B	48) C	60) D	72) B
37) D	49) B	61) C	73) D
38) E	50) C	62) C	74) B
39) B	51) A	63) B	75) D
40) C	52) C	64) B	

Answers to true-false test for zone 1

1) T	5) T	8) F
2) T	6) T	9) T
3) F	7) T	10) T
4) F		

Answers to true-false test for zone 2

1) T	5) F	8) T
2) T	6) T	9) T
3) T	7) T	10) T
4) F		

Answers to true-false test for zone 3

1) F	5) F	8) T
2) F	6) T	9) T
3) F	7) T	10) F
4) T		

CHAPTER 3

Understanding code definitions

THE plumbing code is full of words that are not used on a regular basis by average people. Understanding code definitions is only a matter of memory. If you have a good memory, you will not have any trouble learning the word definitions used in the plumbing code. However, when it comes time to take the exam, the meanings of certain words might escape your memory. What's worse, the exam might be worded in such a way as to confuse or trick you.

While definitions usually are not a major part of a licensing exam, they will appear on the test. Since every wrong answer on the exam counts against you, it is wise to know the word definitions. Some of the words you will already know; others you will have to study. Don't overlook this aspect of the test. Correct answers on the definition questions might mean the difference between a passing or failing test result.

If you open a code book, you will find definitions near the front of the book. There are a lot of them. While many of the words seem simple to understand, it is easy to make a mistake on the day of the exam. Pure memorization of the code book text might not be enough to get you through the test. It has been my experience that many plumbing exams are designed to test more than your memory skills. The test is meant to determine your understanding of the code and definitions. This fact gets some people in trouble.

One of the students who attended my class on code preparation had a very good memory. He could read a page of definitions and practically recite the words verbatim. But he was not as gifted at applying the code. This fact came to light when I gave the class their test on definitions.

As an instructor, I attempt to simulate real-world situations. In regards to the tests that I create, this means the wording in the questions occasionally must be a little tricky. The right questions can pinpoint quickly the people who have memorized the code rather than learned it. This was the case with the student who thought he knew all the definitions by heart. His work on the mock exam showed him that he didn't know the intent of the code. Fortunately, this person found his faults in my classroom rather than at the real exam.

This chapter is going to do for you what I did for the student in my class. It is going to test your understanding of the definitions. I won't bore you by duplicating the definitions as they are presented in the plumbing code. Instead, I'll give you work that will determine just how well you really understand the words and terms. Let's start this test with a multiple-choice exam. The questions in the test might have more than one correct answer. Choose the answer that is most closely related to the code definition.

Multiple-choice exam

1) When you read the code in regards to types of pipe that may be used in a plumbing system, you will often see the abbreviation of ABS used to represent a particular type of pipe. What does ABS stand for?

   A) Anti-Block System
   B) Acrylonitrile-Butadiene-Styrene
   C) American-Butadiene-Styrene

2) When you have a shower valve in a wall that can be accessed with the removal of an access panel, which of the following words best represents the condition of the shower valve?

   A) Accessible
   B) Readily accessible
   C) Partially accessible

3) If you walk into an open basement and observe the gate valve on a water heater, which of the following words best describes the condition of the gate valve?

A) Accessible
B) Readily accessible
C) Partially accessible

4) When there is a physical separation between a discharge pipe and the drain that accepts the discharge, what do you have? Choose the word that best describes the situation.

A) A broken pipe
B) An airbreak
C) An open drain

5) What term best describes the unobstructed vertical distance through the free atmosphere between the lowest opening from a pipe that is conveying waste to an open receptor?

A) Free fall
B) Air break
C) Air gap

6) What term best describes the unobstructed vertical distance through the free atmosphere between the lowest opening from a pipe that is conveying potable water to the flood-level rim of a fixture?

A) Free fall
B) Air break
C) Air gap

7) When something is accepted or acceptable under an applicable standard of the plumbing code, it is said to be:

A) All right
B) Compliant
C) Approved

8) An organization that is primarily established for the purpose of testing to approved standards and whose existence has been approved by the proper authorities is known as:

A) An engineering firm
B) An approved testing agency
C) A certified plumbing company

9) When water enters a potable water pipe from any source other than its intended source, the process is known as:

A) Reverse flow
B) Back flow
C) Cross connection

10) What type of valve should be installed in certain pipes to prevent the reverse flow of drainage substances?

A) Reverse-flow valve
B) Backwater valve
C) Gate valve

11) Which of the following is the definition of a bathroom?

A) A room equipped with a bathtub or shower
B) A room equipped with a toilet and lavatory
C) A room equipped with a toilet, lavatory, and bathing unit

12) What is the device that is present on a boiler that permits the emptying or discharge of sediment?

A) Relief valve
B) Boiler blowoff
C) Aquastat

13) A connection or condition that might result in back flow is known as:

A) A back flow connection
B) A back flow condition
C) A back flow condenser

14) Increased pressure in a system can result in back flow. When back flow is caused by a pressure that is in excess of the supply pressure, the condition is known as:

A) Back-puffing
B) Back-pressure back flow
C) Pressure imbalance

15) A device that is used to prevent back flow into a potable water supply is known as:

A) Backwater valve
B) Backwater preventer
C) Back flow preventer

16) When used, contaminated, or polluted water from a plumbing fixture flows back into the potable water supply due to a negative pressure, the condition might be called back flow or:

A) Reverse flow
B) Back-siphonage
C) Counter-siphonage

17) Any part of a piping system that is not a main, riser, or stack is a(n):

A) Arterial
B) Branch
C) Feed

18) A vent that connects one or more individual vents with a vent stack or stack vent is a:

A) Common vent
B) Wet vent
C) Branch vent

19) The drainage pipe that is the lowest pipe in the drainage system and that receives the discharge from soil, waste, and other drainage pipes within the inside walls of a building and conveys it to a pipe that begins two feet outside the building wall is known as a:

A) Sewer
B) Building sewer
C) Building drain

20) What pipe accepts the discharge from the pipe described in question 19?

A) Sewer
B) Building sewer
C) Building drain

21) What is another name for a water service?

A) Water main
B) Building supply
C) Main inlet

22) A lined excavation in the ground that receives the discharge of a drainage system and that has been designed to retain organic matter and solids while allowing liquids to seep through the bottom and sides is known as:

A) Leech field
B) Septic tank
C) Cesspool

23) A plumbing system that is designed with the waste piping embodying the horizontal wet venting of one or more sinks or floor drains by means of a common waste and vent pipe that is adequately sized to provide free movement of air above the flowline of the drain is known as:

A) A combination waste and vent system
B) An engineered system
C) An illegal system

24) For the purposes of the plumbing code, combustible construction will ignite and burn at a temperature of:

A) 500 degrees Fahrenheit
B) 1000 degrees Fahrenheit
C) 1392 degrees Fahrenheit

25) An impairment of the quality of potable water that creates a hazard to public health through poisoning or the spread of disease is known as:

A) Contamination
B) Back flow
C) Risk danger

26) A vent that is run vertically and extends as a continuation of the drain it serves is known as a:

A) Common vent
B) Vent stack
C) Continuous vent

27) A waste arrangement that connects multiple compartments of a set of fixtures, such as a double-bowl kitchen sink, to a single trap is known as:

A) An island waste
B) A collective waste
C) A continuous waste

28) A connection between a potable water supply and any fixture or device where non-potable water can enter the potable water supply is known as:

A) A cross connection
B) A reverse connection
C) A back flow connection

29) The minimum elevation at which a device can be installed above the flood-level rim of a fixture or receptacle served by a back flow preventer is known as the:

A) Cut-off point
B) Critical level
C) Installation level

30) When a pipe is measured along the centerline of the pipe and fittings used with the pipe, the measurement determines the:

A) Combined length
B) Developed length
C) Overall length

31) A pipe that conveys waste or water-borne wastes in a building drainage system is known as:

A) A branch
B) A stack
C) A drain

32) Liquid and water-borne wastes that are derived from ordinary living processes and are free from industrial wastes or other wastes that might require special treatment other than the simple discharging into a public sewer is known as:

A) Common sewage
B) Safe sewage
C) Domestic sewage

33) A plumbing system or any part of a plumbing system that was installed prior to the effective date of the plumbing code is deemed to be:

A) Approved work
B) Existing work
C) Exempt work

34) A water-supply pipe between a fixture-supply pipe and a water-distribution pipe is known as a:

A) Branch fixture
B) Fixture branch
C) Fixture supply

35) The drainage piping that extends from the trap of a fixture to a point of connection with another drain pipe is called a:

A) Trap arm
B) Horizontal stack
C) Fixture drain

36) A water supply pipe that connects a fixture with a fixture branch is called a:

A) Durham fitting
B) Fixture supply
C) Fixture feed

37) The top edge of a fixture or receptacle from which water can overflow is called a:

A) Critical level
B) Flood-level rim
C) Flood surface

38) The slope or fall of a drainage pipe is normally called either the pitch of the pipe or the _____ of the pipe. Fill in the blank.

A) Elevation
B) Drop
C) Grade

39) A device designed to retain grease from up to four fixtures is called a:

A) Grease trap
B) Grease interceptor
C) Grease diverter

40) A device with at least a 750 gallon capacity that is meant to serve one or more fixtures in preventing grease from entering the building drain and which is remotely located is called a:

A) Grease trap
B) Grease interceptor
C) Grease diverter

41) Another name for a house drain is:

A) Building sewer
B) Building drain
C) Building trap

42) Another name for a house sewer is:

A) Building sewer
B) Building drain
C) Building trap

43) A horizontal pipe must be installed horizontally and make no angles that are greater than _____ degrees with the horizontal. Fill in the blank.

A) 22½
B) 45
C) 60

44) A pipe that is installed to vent a fixture trap and which connects with the vent system above the fixture or terminates into the open air is called a:

A) Individual vent
B) Vacuum vent
C) Wet vent

45) Any liquid or water-borne waste from industrial or commercial processes, except domestic sewage is known as:

A) Industrial waste
B) Commercial waste
C) Either A or B

46) A pipe that does not connect directly with the drainage system, but conveys liquid wastes by discharging them into a plumbing fixture that is connected to the drainage system is called a:

A) Air gap
B) Open-air waste
C) Indirect waste

47) When a condition that is contrary to sanitary principles or is injurious to health exists, that condition is known as:

A) Unsanitary
B) Insanitary
C) Detrimental

48) Discharge that does not contain fecal matter is called:

A) Chemical waste
B) Liquid waste
C) Modified sewage

49) When a plumbing fixture is not supplied with water sufficient to flush it and maintain a clean condition, the fixture is deemed to be:

A) Unsanitary
B) Insanitary
C) Detrimental

50) Any trap that does not maintain a proper trap seal is deemed to be:

A) Unsanitary
B) Insanitary
C) Detrimental

51) A vent that connects a horizontal branch or fixture drain with a stack vent of the originating waste or soil stack is called a:

A) Relief vent
B) Bombay vent
C) Loop vent

52) The principal artery of a venting system to which vent branches might be connected is called a:

A) Vent stack
B) Stack vent
C) Main vent

53) A combination of elbows or bends that brings one section of pipe out of line but into a line parallel with the other section of the pipe is known as:

A) An offset
B) A drainage turn
C) A crown vent

54) What is the abbreviation for polybutylene?

A) PB
B) PE
C) PBL

55) What is the abbreviation for polyethylene?

A) CPVC
B) PE
C) PEL

56) Potable water is water that may be used for which of the following:

A) Domestic purposes
B) Culinary purposes
C) Both A and B

57) Another name for drinking water is:

A) Potable water
B) Gray water
C) Chlorinated water

58) A sewer that receives the discharge from more than one building drain and conveys it to a point of disposal is called a:

A) Common sewer
B) Elevated sewer
C) Private sewer

59) The letters PVC are an abbreviation for:

A) Polyvinyl chloride
C) Polyvinyl calcium
B) Polyvinyl clusters

60) A plumbing fixture or device that has been designed and approved to receive the discharge from indirect waste pipes is called:

A) A dump station
C) An airbreak
B) A receptor

61) A vent whose primary function is to provide circulation of air between drainage and vent systems or to act as an auxiliary vent on a specially designed system is known as a:

A) Relief vent
C) Vent stack
B) Loop vent

62) The unobstructed open edge of a fixture is known as the:

A) Rim
C) Edge
B) Overflow

63) A lined excavation in the ground that receives the discharge of a septic tank so designed as to permit the effluent from the tank to seep through its bottom and sides is known as a:

A) Leech field
C) Seepage pit
B) Cesspool

64) Any liquid waste that contains animal or vegetable matter in suspension or solution is known as:

A) Liquid waste
C) Domestic waste
B) Sewage

65) An adjustable tubing connection that consists of a compression nut, friction ring, and compression washer designed to fit a threaded adapter fitting or a standard taper pipe thread is called a:

A) Slip joint
C) Compression joint
B) Temporary connection

66) The vertical main of a system of soil, waste, or vent piping that extends through one of more stories is called a:

A) Stack
B) Riser
C) Stack vent

67) A soldered joint is one obtained by joining metal parts with metallic mixtures or alloys that melt at a temperature above 300 degrees and below _____ degrees Fahrenheit. Fill in the blank.

A) 500
B) 800
C) 1000

68) A water supply pipe that extends vertically for one full story, or more, is called a:

A) Stack
B) Riser
C) Main

69) Wastes that require some special method of handling are known as:

A) Chemical wastes
B) Special wastes
C) Industrial wastes

70) A pipe that is an extension of a soil or waste stack above the highest horizontal drain is called a:

A) Stack vent
B) Vent stack
C) Main stack

71) A pipe that connects upward from a soil or waste stack to a vent stack to prevent pressure changes in the stacks is known as:

A) A relief vent
B) A pressure-reducing vent
C) A yoke vent

72) An approved tank or pit that collects sewage or liquid waste and that is located below the normal grade of the gravity plumbing system and that requires mechanical means to be emptied is called a:

A) Cesspool
B) Sewage well
C) Sump

73) The portion of a fixture drain that runs between the trap of the fixture and the vent for the fixture is known as a:

A) Fixture drain
B) Trap arm
C) Wet vent

74) The maximum vertical depth of liquid that a trap will retain, measured between the crown weir and the top of the dip of the trap is known as:

A) The trap seal
B) The trap level
C) The critical level

75) A vertical vent pipe installed primarily to provide air circulation to and from any part of the drainage system is known as a:

A) Vent stack
B) Stack vent
C) Common vent

Lay your pencil down and rest your eyes. You've just completed the multiple-choice exam on definitions. Don't check your answers yet, you still have a true-false test to take.

How do you think you did on this part of the testing? Did some of the questions pose problems for you? Unless you possess an uncommon grasp of code definitions, you probably were confused by some of the potential answers. Don't feel bad, you were supposed to be confused. The purpose of this book is to make you think and to expose your weak points. By doing this, you will be prepared for your exam. Take a short break, if you like, and then move on to the true-false questions that pertain to your region.

True-false test for zone 1

1) A pipe that conveys potable water from a building supply to a plumbing fixture is a water distribution pipe.

True False

2) A welded joint is any joint obtained by the joining of metal parts in the plastic molten state.

True False

3) A waste pipe is a pipe that conveys liquid waste and suspended fecal matter.

True False

4) The purpose of a trap is to prevent the back passage of air without materially affecting the flow of sewage or waste water through it.

True False

5) Wet vents also serve as drains.

True False

6) A soil pipe is a pipe that conveys the waste of sump pumps, laundry tubs, and lavatories.

True False

7) A sewer that is directly controlled by public authority and that serves as a common sewer is known as a public sewer.

True False

8) A cylindrical conduit that conforms to the particular dimensions commonly known as pipe size is an approved Durham fitting.

True False

9) Inadequate or unsafe water supply or sewage disposal systems are considered to be a nuisance.

True False

10) Grease interceptor is just another name for a grease trap.

True False

True-false test for zone 2

1) A valve that is designed to be a one-way valve to allow air to enter a plumbing drainage system when negative pressures develop in the piping is called an air-admittance valve.

True False

2) A pressure balancing valve that senses incoming hot and cold water pressures and compensates for fluctuations in either the hot or cold inlet to stabilize the temperature of water being produced from the valve is called an anti-scald valve.

True False

3) A plumbing receptacle that is designed to collect storm water in an open area is called an indirect waste.

True False

4) A group of two or more similar adjacent fixtures that discharge into a common horizontal waste or soil pipe is called a continuous waste.

True False

5) A device installed in a building drain to prevent the circulation of air between the drainage system of the building and the building sewer is called a drum trap.

True False

6) A branch vent that serves two or more traps and extends from in front of the last fixture connection of a horizontal branch to a vent stack is called a combination vent.

True False

7) Two changes in direction installed in succession are called a double offset.

True False

8) A Durham system is one where all soil or waste systems are made of rigid, threaded pipe and recessed drainage fittings.

True False

9) Lead-free pipe and fittings contain no more than 8 percent lead.

True False

10) Lead-free solder and flux contains no more than 0.2 percent lead.

True False

True-false test for zone 3

1) A bathroom group, as defined by the code, consists of a water closet, bidet, lavatory, and bathtub or shower located together on the same floor level.

   True False

2) A portion of a drainage system that does not drain by gravity into a building sewer is called a building subdrain.

   True False

3) A branch leading from a soil, waste, or vent pipe and terminating at a developed length of 1 foot or less that is capped or plugged is called a dead end.

   True False

4) An emergency floor drain is a drain that receives the discharge of a fixture through an indirect waste.

   True False

5) A sewer ejector is a device used to lift sewage by entraining the sewage in a high-velocity jet of steam, air, or water.

   True False

6) To be classified as a swimming pool, a body of water for swimming, diving, or recreational bathing must have a depth of at least 3 feet.

   True False

7) An exterior drainage pipe that carries storm water from a roof or gutter drain is known as a leader.

   True False

8) Sump pumps are not intended to remove raw sewage.

   True False

9) A pipe located inside a building to carry storm water from the roof to a storm or combined building drain is known as a conductor.

True False

10) A vent that extends from a pneumatic sewage ejector and terminates separately to the open air is called an ejector vent.

True False

You have now completed all of the tests on definitions. Consult the correct answers at the end of this chapter and compare them with your answers. Take note of any errors made and review your code book to learn why the correct answers are what they are.

As I said in the early part of this chapter, knowing what a word or term means is important. It is important when you take your test and it's important when you work in the field. Take the time to understand the intent of the code. Don't just memorize the text for the test. Study the code until you have a solid understanding of what it means and how it should be applied.

Answers to multiple-choice exam

1) B	16) B	31) C	46) C
2) A	17) B	32) C	47) B
3) B	18) C	33) B	48) B
4) B	19) C	34) B	49) B
5) C	20) B	35) C	50) B
6) C	21) B	36) B	51) C
7) C	22) C	37) B	52) C
8) B	23) A	38) C	53) A
9) B	24) C	39) A	54) A
10) B	25) A	40) B	55) B
11) A	26) C	41) B	56) C
12) B	27) C	42) A	57) A
13) A	28) A	43) B	58) C
14) B	29) B	44) A	59) A
15) C	30) B	45) A	60) B

61) A	65) A	69) B	73) B
62) A	66) A	70) A	74) A
63) C	67) B	71) C	75) A
64) B	68) B	72) C	

Answers to true-false test for zone 1

1) T	5) T	8) F
2) T	6) F	9) T
3) F	7) T	10) F
4) T		

Answers to true-false test for zone 2

1) T	5) F	8) T
2) T	6) F	9) T
3) F	7) T	10) T
4) F		

Answers to true-false test for zone 3

1) T	5) T	8) T
2) T	6) F	9) T
3) F	7) T	10) F
4) F		

CHAPTER 4

General code regulations you must know

IN addition to the many specific aspects of the plumbing code, there is a fair number of general code regulations you must know. These elements of the code are just as important as any other when it comes to taking a licensing exam. Not only are the general regulations required reading for the test, many of them are used frequently in the field. For example, the general regulations of the code dictate how the testing of a plumbing system must be done.

If you are not already familiar with the general regulations, you will find that there are many of them that will be used on a regular basis as you go about your plumbing work. Before we jump right into another mock exam, let's talk for a few moments about general regulations.

You might say that the general regulations are the foundation on which the plumbing code is built. Putting it in terms of construction, you might say that the general regulations are the footing for a structure. If you know anything about construction, you know how critical a good footing is to the construction of a building. Just as a footing is of paramount importance in construction, the general regulations of the plumbing code are equally important in the application and compliance of the code.

What types of requirements do the general regulations cover? The list of subjects that pertain to the general regulations is not particularly long, but it is important. The general regulations deal with health and safety. As a plumber, you might think health and safety pertains only to potable water and sewage. In reality, the general regulations go much deeper than that. Although some of the regulations might not seem specific to plumbing, they are related to work that is done in conjunction with the installation of plumbing. Let's talk about one such example.

When you read the general regulations, you will notice provisions that deal with the cutting and notching of structural members. You might think this information applies more to carpenters than plumbers, but plumbers do a lot of cutting and notching. When a floor joist blocks the path of a drain for a closet flange, something must be done. As the plumber on the job, you might have to decide on the right plan of action. Should you notch out the top of the joist, or should you cut the joist completely away and install headers on the severed ends?

Suppose you are running a horizontal drain through floor joists and cannot keep the pipe in the center of the joists? Since the drain must be graded properly, you might find that the holes are close to the bottom of the joist. How close to the bottom of a joist can you drill a hole without causing structural or code problems? If you are in charge of plumbing the job, you'd better know. Drilling holes too close to the bottom of joists can result in some expensive and time-consuming repairs for which you might be held responsible. As you can see, there are important parts of the general regulations that you might, at first, take for granted.

Let's look at another example. Assume that you are in charge of installing the building sewer. Due to the location of the main sewer, the building sewer must run parallel to the footing of the structure. There is a backhoe operator on the job to dig the trench for you, but you must explain to the operator where and how to dig the trench. Are you aware that there is a rule in the general regulations that pertains to this type of situation? Well, there is. The rule requires that the trench shall not extend below the line of a 45-degree angle to the loadbearing plane of the footing. The reason for this rule is fairly obvious; a more severe trench could weaken the structural integrity of the footing. Would you want to be responsible for having a trench dug that undermined a foundation? Of course you wouldn't, and if you learn the general regulations, you can avoid such problems.

I've taken this brief detour from the sample test to impress upon you the importance of the general regulations. I've seen a lot of plumbers, and would-be plumbers pay little attention to most of the general regulations. They assume the rules are not going to affect them. It's true that you might never use some of the rules found in the general regulations, but you never know which rule will come out of left field when you least expect it. You must learn the general regulations for your exam. Learn the general regulations to enjoy a safer and more productive time in the field. With that said, let's get on with the multiple-choice exam.

Multiple-choice exam

1) It is common practice to make all measurements, such as the distance from one fixture drain to another, to what portion of a pipe?

   A) The left side
   B) The right side
   C) The center
   D) The side closest to the adjoining fixture

2) When a health or safety hazard exists on a premises due to an existing plumbing system or the lack of one, who is responsible to abate such a nuisance?

   A) A code officer
   B) A master plumber
   C) The property owner or an authorized agent of the same
   D) A journeyman plumber

3) A plumbing fixture that discharges only waste, such as a kitchen sink, can be drained into which of the following:

   A) A sanitary drainage system
   B) A French drain
   C) A gravel-lined pit
   D) Either A or B

4) When altering a structure for the installation of plumbing, the job must be left in what condition?

   A) A safe and nonhazardous condition
   B) A condition equal to the way it was found

C) A neat and clean condition
D) A condition that is satisfactory to the property owner

5) Trenching that runs parallel to the footing of a structure must not extend below the line of a _____ to the loadbearing plane of the footing. Fill in the blank.

A) 22½-degree angle
B) 45-degree angle
C) 60-degree angle
D) 90-degree angle

6) Before waste products from manufacturing or industrial operations can be dumped into a public sewer, someone must determine if the waste will harm the sewer or sewage treatment facilities. Who has the authority to make this determination?

A) A master plumber
B) A code official
C) An authority authorized for such work
D) Either B or C

7) Strainer plates on drain inlets shall be made with openings that are not larger than _____ in least dimension. Fill in the blank.

A) ¼ inch
B) ½ inch
C) ¾ inch
D) 1 inch

8) What type of facilities must be provided for workers on construction sites?

A) Toilet facilities
B) Rest facilities
C) Recreational facilities
D) None of the above

9) Most plumbing is prohibited from being installed in elevator shafts and elevator equipment rooms. There are, however, some exceptions. Which of the following are an exception to the rule?

A) Floor drains
B) Sumps and sump pumps
C) Both A and B
D) None of the above

10) Meter boxes are required to be constructed in such a way that unwanted creatures cannot enter them. What creature, in specific, does the code refer to?

A) Snakes
B) Cats
C) Rats
D) Spiders

11) When backfilling a trench, the fill material should be installed in layers. How thick should each layer be?

A) 4 inches
B) 6 inches
C) 8 inches
D) 12 inches

12) Backfilling requires the use of loose earth that is free from which of the following:

A) Rocks
B) Broken concrete
C) Frozen chunks
D) All of the above

13) When rock is encountered during trenching for a pipe, the rock must be removed to a level _____ below the grade line of the trench. Fill in the blank.

A) 2 inches
B) 3 inches
C) 4 inches
D) 6 inches

14) Water closets that are installed for public use, and that are not installed in a single-occupant toilet room equipped with a lockable door, must occupy a separate compartment. These compartments must contain which of the following:

A) A door
B) Walls or partitions
C) A handrail
D) Both A and B

15) Piping installed as part of a plumbing system must be installed to prevent which of the following:

A) Strains
B) Stresses
C) Both A and B
D) None of the above

16) It might be necessary to protect plumbing pipes from which of the following:

A) Expansion
B) Contraction
C) Structural settlement
D) All of the above

17) Pipes that pass under or through a wall must be protected from:

A) Rats
B) Condensation
C) Breakage
D) All of the above

18) Any pipe that passes under a footing or through a foundation must be protected by a relieving arch or a ______. Fill in the blank.

A) Sleeve
B) Counterbalance
C) Backwater valve
D) Any of the above

19) When a sleeve is used for pipe protection, the sleeve must have a diameter at least ____ pipe sizes larger than the pipe being protected. Fill in the blank.

A) 2
B) 3
C) Either A or B
D) None of the above

20) There are times when pipes passing through concrete must be protected from corrosion, which of the following is a common form of protection in these circumstances?

A) A protective coating
B) A protective wrapping
C) Both A and B
D) None of the above

21) Which of the following types of pipe are not required to be protected from corrosion caused by passing through concrete?

A) Copper
B) PVC
C) ABS
D) Both B and C

22) If soft soil conditions are encountered when trenching for a pipe installation and solid ground is not readily available, it will be necessary to over excavate the trench and backfill to the desired bed level to assure a solid base for bedding the pipe. When over excavating is necessary, how far does the code require the trench depth to be extended before backfilling with a solid fill?

A) A minimum of one pipe size
B) A minimum of two pipe sizes
C) 4 inches
D) 6 inches

23) Which of the following materials is suitable as a stabilization backfill material when a trench for piping has been over excavated?

A) Gravel
B) Crushed stone
C) Concrete
D) All of the above

24) When a trench is backfilled, at what intervals must the fill material be compacted?

A) Every 4 inches
B) Every 6 inches
C) Every 8 inches
D) Every 12 inches

25) Which of the following types of pipe do not require the installation of nail plates to protect them?

A) PVC
B) ABS
C) Copper
D) Cast iron

26) Which of the following types of pipe do not require the installation of nail plates to protect them?

A) PVC
B) PB
C) Copper
D) Galvanized steel

27) If a copper pipe is installed within _____ from the edge of a stud, where it might be hit by a nail or a screw, the pipe must be protected with a nail plate. Fill in the blank.

A) ¾ of an inch
B) 1 inch
C) 1¼ inches
D) 1½ inches

28) What is the minimum thickness required for the construction of a nail plate?

A) 1⁄16 inch
B) ⅛ inch
C) ¼ inch
D) ½ inch

29) When a nail plate is installed to protect a pipe that is passing through a sole plate, the nail plate must extend how far above the sole plate?

A) ½ inch
B) 1 inch
C) 1½ inches
D) 2 inches

30) When a nail plate is installed to protect a pipe that is passing through a top plate, the nail plate must extend how far below the top plate?

A) ½ inch
B) 1 inch
C) 1½ inches
D) 2 inches

31) Pipes can be protected from freezing in unheated areas with the use of which of the following:

A) Insulation
B) Heat
C) Either A or B
D) None of the above

32) At what intervals must buried pipe be supported?

A) 4-foot intervals
B) 6-foot intervals
C) 10-foot intervals
D) For its entire length

33) Pipes in trenches may not be covered with backfill until ____. Fill in the blank.

A) The pipe has been tested and approved by a code official
B) The backfill material is dry
C) The pipe has been tested for stress
D) All of the above

34) When drilling holes in joists for the passage of pipes, the holes must not be drilled closer than _____ from the top or bottom of the joist. Fill in the blank.

A) 1 inch
B) 1½ inches
C) 2 inches
D) 2¼ inches

35) Notches on the ends of joists must not exceed _____ of the depth of the joist. Fill in the blank.

A) one eighth
B) one fourth
C) one half
D) three quarters

36) Any hole drilled in a joist must not have a diameter that exceeds _____ the depth of the joist. Fill in the blank.

A) one eighth
B) one third
C) one fourth
D) one half

37) Notches cut in the tops or bottoms of joists for the passage of pipes must not be cut to a depth that exceeds _____ the depth of the joist. Fill in the blank.

A) one eighth
B) one third
C) one fourth
D) one sixth

38) Studs in exterior walls and bearing partitions can be cut or notched, but the cut or notch must not exceed _____ of the stud width. Fill in the blank.

A) 15 percent
B) 20 percent
C) 25 percent
D) 40 percent

39) Studs in nonbearing partitions can be cut or notched, but the cut or notch must not exceed _____ of the stud width. Fill in the blank.

A) 15 percent
B) 20 percent
C) 25 percent
D) 40 percent

40) A hole can be drilled in any stud as long as the hole diameter does not exceed _____ of the stud width. Fill in the blank.

A) 25 percent
B) 30 percent
C) 40 percent
D) 60 percent

41) A hole can be drilled in a stud in a nonbearing partition as long as the hole diameter does not exceed _____ of the stud width. Fill in the blank.

A) 25 percent
B) 30 percent
C) 40 percent
D) 60 percent

42) Holes drilled through studs should not be placed closer than _____ to the edge of the stud. Fill in the blank.

A) ¼ of an inch
B) ⅜ of an inch
C) ⅝ of an inch
D) ¾ of an inch

43) What must be done with the annular space between a pipe and a sleeve that protects the pipe?

A) It must be left open
B) It must be ventilated
C) It must be protected from freezing
D) It must be filled or tightly caulked

44) The minimum depth for burying a water service pipe is either 12 inches or _____ inches below the local frost line, whichever is deeper. Fill in the blank.

A) 4
B) 5
C) 6
D) 8

45) Exterior openings made for plumbing must be made _____. Fill in the blank.

A) Rodent proof
B) Watertight
C) Both A and B
D) None of the above

46) Plumbing installations must not interfere with which of the following:

A) Windows
B) Doors
C) Egress openings
D) All of the above

47) When water is used to test underground plumbing drains and vents, the use of a standpipe is required for the entry and observation of water used for the test. How tall is this standpipe required to be?

A) 4 feet
B) 5 feet
C) 8 feet
D) 10 feet

48) How long is water required to remain in a drainage system prior to an inspection of the joints?

A) 10 minutes
B) 15 minutes
C) 20 minutes
D) 30 minutes

49) If air is used to test a drain, waste, and vent system, what is the required test pressure?

A) 5 pounds per square inch
B) 10 pounds per square inch
C) 15 pounds per square inch
D) 18 pounds per square inch

50) For what period of time must a drain, waste, and vent system hold its test pressure to prove that there are no leaks in the system?

A) 10 minutes
B) 15 minutes
C) 20 minutes
D) 30 minutes

51) When a peppermint test is conducted on a drain, waste, and vent system, how much oil of peppermint must be introduced into each stack?

A) 2 ounces
B) 1 ounce for each pipe size of the stack
C) 3 ounces
D) Either A or B

52) If water is used to test water distribution piping, the water used in the test must be obtained from _____. Fill in the blank.

A) A well
B) A sanitary drum
C) A potable source
D) Licensed testing agency

53) The minimum depth for burying a sewer is _____. Fill in the blank.

A) 6 inches
B) 12 inches
C) 18 inches
D) 24 inches

54) When testing a sewer installation for leaks with water, the pipe must be fitted with a standpipe for the introduction and observation of test water. What is the minimum height requirement for such a standpipe?

A) 4 feet
B) 6 feet
C) 8 feet
D) 10 feet

55) Who is responsible for testing new plumbing installations?

A) A master plumber
B) The property owner or an authorized agent of the same
C) The permit holder
D) A code official

You have just completed the multiple-choice exam that pertains to general regulations. If you answered all of these questions correctly, you certainly have a good grasp of the general regulations applied in the plumbing code. Before you check your answers, however, take the true-false test that covers your region. After completing the true-false test you can check all of your answers.

True-false test for zone 1

1) The plumbing code prohibits the threading of cast-iron pipe.
   True False

2) Drainage and vent piping may not be drilled and tapped for the purpose of making a connection.
   True False

3) Side-inlet closet bends may be used to receive the discharge from a lavatory.
   True False

4) Pipe trenches shall be filled in thin layers until there is a minimum of 12 inches of cover over the pipe.
   True False

5) Inspection requests must be made at least 24 hours in advance of the time the inspection is desired.
   True False

6) Water distribution pipes must be tested with water at the working pressure of the system or with air at a pressure of 50 pounds per square inch.
   True False

7) Shower receptors must be tested by running water through the shower head for at least 5 minutes.
   True False

8) If a mercury gauge is used to test a drain, waste, and vent system, the system must be filled with air that will suspend and balance a column of mercury 10 inches in height.
   True False

9) If a building is moved from one foundation to another, it is required that the entire plumbing system be tested in accordance with code requirements.

True False

10) Air may not be used to test a sewer for leaks.

True False

True-false test for zone 2

1) Water distribution pipes must be tested at a pressure that is at least 25 pounds per square inch higher than the normal working pressure.

True False

2) In certain instances, a code official may require a final plumbing inspection to be done with the use of either oil of peppermint or a pungent, thick smoke.

True False

3) All joints at the roof and around vent pipes must be made watertight with the use of either lead, copper, galvanized steel, aluminum, plastic, or other approved materials.

True False

4) Under no conditions can a water service and building sewer be installed in the same trench.

True False

5) Holes with diameters larger than 2$\frac{9}{16}$ inches are prohibited in ceiling joists.

True False

6) A side-inlet quarter bend can only be used as a vent when the inlet is placed in a horizontal position.

True False

7) Short sweeps that are not less than 3 inches in diameter can be used in soil and wastes lines where the change in direction is from vertical to horizontal or from horizontal to vertical.

True False

8) Double sanitary tees may be used in drainage lines only where the direction of flow is horizontal.

True False

9) Saddle-type fittings may not be used with a drainage system.

True False

10) Running threads may be used in vent systems, but not in drainage systems.

True False

True-false test for zone 3

1) It is a violation of the plumbing code to deposit ashes, cinders, rags, flammable substances, poisonous substances, or explosive liquids in a drainage system.

True False

2) Plumbing systems installed in buildings located in a flood-hazard zone must be installed to withstand the effects of buoyancy during the occurrence of flooding to the base flood elevation.

True False

3) The annular space between a pipe and a pipe sleeve must be filled with an expanding foam.

True False

4) Pipe in trenches shall not be supported on blocks to grade.

True False

5) A concrete foundation used to provide a solid base for pipe installation in a trench must be poured over a layer of gravel and wire mesh if the trench has a soft loadbearing ability.

True False

6) All types of pipe must be protected from possible penetration by nails or screws with the use of nail plates.

True False

7) Floor drains may be installed at the base of an elevator shaft.

True False

8) Non-sewer toilet facilities are an acceptable means by which to provide toilet facilities for construction workers.

True False

9) Any opening for a water service pipe must be sealed and protected by the installation of approved metal collars.

True False

10) The use of tunneling for the installation of pipe shall only be done by licensed master plumbers.

True False

Well, you've completed all the tests on general regulations. Turn to the answer section at the end of this chapter and check your work. Note and study any questions that you answered incorrectly. Consult your code book to find out why you answered the question incorrectly. Once you've done that, turn to chapter 5 and take the tests that pertain to plumbing materials. I'll meet you there.

Answers to multiple-choice exam

1) C	5) B	9) C	13) B
2) C	6) D	10) C	14) D
3) A	7) B	11) B	15) C
4) A	8) A	12) D	16) D

17) C	27) C	37) D	47) D
18) A	28) A	38) C	48) B
19) A	29) D	39) D	49) A
20) C	30) D	40) C	50) B
21) D	31) C	41) D	51) A
22) B	32) D	42) C	52) C
23) D	33) A	43) D	53) B
24) B	34) C	44) C	54) D
25) D	35) B	45) C	55) C
26) D	36) B	46) D	

Answers to true-false test for zone 1

1) T	5) T	8) T
2) T	6) T	9) T
3) F	7) F	10) F
4) T		

Answers to true-false test for zone 2

1) T	5) F	8) F
2) T	6) F	9) T
3) T	7) T	10) F
4) F		

Answers to true-false test for zone 3

1) T	5) F	8) T
2) T	6) F	9) T
3) F	7) T	10) F
4) T		

CHAPTER 5

Choosing the proper plumbing materials

WITH the wide variety of materials available for different types of plumbing work, keeping all the code requirements clear can be troublesome. Not only do you have to know what types of materials are acceptable for certain types of plumbing applications, you should be able to determine which of the approved materials is the best choice for each job. For example, both galvanized steel pipe and Schedule-40 plastic pipe are approved materials for most types of drainage. Given the job of choosing which of these two materials to use for a common, residential sink drain, which would you choose?

Either galvanized steel or Schedule-40 plastic pipe will earn you a correct answer on your plumbing exam. But, in the field, Schedule-40 plastic pipe would be the best choice. Galvanized steel pipe tends to rust. The rust creates rough spots on interior sections of the pipe. In time, these rough sections act as a trap for hair, grease, and other unwanted debris. Over a little more time, the pipe begins to close up, until eventually the interior of the pipe is obstructed. When this happens, water cannot drain past the clog, and the waste water backs up into the sink.

The threaded joints of galvanized pipe are another trouble spot. Since the wall thickness of the pipe is minimal where the threads have been cut, leaks often occur around the threaded sections. This is a result of corrosion.

While plastic pipes occasionally become clogged, they do so less frequently than galvanized steel pipes. It is possible for a joint connection between plastic pipes to go bad and pull apart, but there are no threads to create weak spots. Since plastic is resistive to most corrosive action and rust, it does not deteriorate like steel pipe. These factors combine to make plastic pipe a better all-round choice for drainage pipe.

When you sit for the plumbing exam, you will sometimes see questions that could have more than one right answer. In these situations, choose the best answer. If the example I've just given you were such a question, the correct answer would be Schedule-40 plastic pipe. Even though either pipe is an approved material, plastic pipe is the better choice.

This chapter is going to test your knowledge of approved plumbing materials. Although many of the questions will have only one right answer, some might have more than one. In these cases, select the best available answer. Review Tables 5-1 through 5-6 at the end of this chapter for approved materials.

The first test is a multiple-choice exam. Read the questions carefully. Like the real exam, some of them are worded to catch you off guard. After reading the question, evaluate the possible answers and make your selection.

When you have completed the multiple-choice questions, move on to the true-false test for your region. Do this before you check your score on the multiple-choice exam. Again, read the questions carefully, and look out for trick wording.

Once you have completed all of the test questions, go to the end of the chapter to find the correct answers. Grade your tests and see how you made out.

Real plumbing exams are timed, so don't linger too long on any one question. Move through the test as if you were sitting for that all-important master's license.

After you grade your test, spend some time reviewing the questions you answered incorrectly. Figure out why you made the wrong choice, and commit the rule to memory. There will be no second chance on the day of the actual exam. Unless you are willing to wait for a new test session, pay an additional testing fee, and work for less money than a master plumber makes, you must pass your exam on the first try. This book can help you prepare to do just that, but you have to put forth the effort to learn, and you must take these sample exams seriously. Now, let's get on with the tests.

Multiple-choice exam

1) Which of the following, when used in a plumbing system, must have a cast, embossed, stamped or indelibly marked symbol of the manufacturer?

   A) Pipe
   B) Pipe fittings
   C) Traps
   D) Fixtures
   E) All of the above
   F) None of the above

2) Water service pipe must be resistant to which of the following:

   A) Corrosive action
   B) Degrading action
   C) Both A & B
   D) None of the above

3) Water distribution pipe must be resistant to which of the following:

   A) Corrosive action
   B) Degrading action
   C) Both A & B
   D) None of the above

4) When detrimental conditions are suspected to exist in an area where a water service is to be installed, what must be done?

   A) Copper pipe must be used
   B) Polybutylene pipe must be used
   C) A chemical analysis of the soil must be made
   D) Any of the above

5) What is the maximum percentage of lead allowed in pipe, fittings, valves, and faucets used in a plumbing system?

   A) 0.2 percent
   B) 2 percent
   C) 4 percent
   D) 8 percent

6) When pipe or tubing is used underground as a water service, and is installed outside of a building, it must have a minimum pressure rating of which of the following, when rated at 73.4 degrees Fahrenheit?

A) 80 psi
B) 100 psi
C) 140 psi
D) 160 psi

7) When plastic piping is used as a water service, the plastic piping may not extend farther than _____ within a building.

A) 5 feet
B) 8 feet
C) 10 feet
D) None of the above

8) Which of the following materials may be used as a water service?

A) Copper pipe
B) Polybutylene pipe
C) Lead pipe
D) A & B
E) All of the above

9) Galvanized steel pipe may be used for which of the following purposes?

A) Water service
B) Water distribution
C) Above-ground vents
D) A & B
E) None of the above

10) Pipe or tubing used to convey hot water must have a minimum pressure rating of _____ when rated at 180 degrees Fahrenheit.

A) 60 psi
B) 80 psi
C) 100 psi
D) 125 psi

11) A building sewer pipe may be made of which of the following materials?

A) Vitrified clay pipe
B) Polybutylene pipe
C) Polyethylene pipe
D) Any of the above

12) If DWV copper is used as a building sewer material, what "Type" of copper must it be?

A) Type M
B) Type L
C) Type K
D) Any of the above
E) A & C
F) B & C

13) If DWV copper is used as an underground building drainage or vent material, what "Type" of copper must it be?

A) Type M
B) Type L
C) Type K
D) Any of the above
E) A & C
F) B & C

14) Can a building sewer pipe be installed in a trench where a water service is installed?

A) No, never
B) Yes, always
C) Yes, under certain circumstances
D) Yes, only if the sewer is made of cast iron

15) Materials used to vent or drain chemical waste must be:

A) Made of copper
B) Made of cast iron
C) Resistant to corrosion and degradation for the chemicals involved
D) Made of stainless steel

16) Pipe that is open-jointed, horizontally split, or perforated may be used for which of the following purposes:

A) Sanitary vents
B) Sanitary drains
C) Sub-soil drains
D) Any of the above
E) None of the above

17) Threaded drainage pipe fittings shall be of which of the following types?

A) Solvent-weld type
B) Exposed-drainage type
C) Recessed-drainage type
D) None of the above

18) Fittings must be made in a way that the flow of water through them is not retarded or obstructed by which of the following:

A) Ledges
B) Shoulders
C) Reductions
D) All of the above
E) None of the above

19) Pipe fittings may be made of which of the following materials:

A) Cast iron
B) Copper
C) Plastic
E) A & B only
D) All of the above

20) If a lead bend or trap is used, what is the minimum thickness required for the wall?

A) ⅛ inch
C) ½ inch
B) ¼ inch
D) None of the above

21) A copper or tubular brass trap must have a thickness rated at which of the following:

A) 15 gauge
C) 20 gauge
B) 17 gauge
D) 22 gauge

22) A copper or tubular brass tail piece must have a thickness rated at which of the following:

A) 15 gauge
C) 20 gauge
B) 17 gauge
D) 22 gauge

23) Sheet lead, when used to flash vent pipes, shall weigh how much per square foot?

A) 2 pounds
C) 4 pounds
B) 3 pounds
D) Any of the above

24) Sheet lead, when used for general use, shall weigh how much per square foot?

A) 2 pounds
C) 4 pounds
B) 3 pounds
D) Any of the above

25) Sheet lead, when prefabricated for use in flashing vent pipes, may weigh as little as:

A) 2 pounds
C) 3 pounds
B) 2½ pounds
D) 3½ pounds

26) Sheet copper, when used for general use, shall weight how much per square foot?

A) 8 ounces
C) 12 ounces
B) 10 ounces
D) 14 ounces

27) Sheet copper, when used to flash vent pipes, shall weigh how much per square foot?

A) 8 ounces
B) 10 ounces
C) 12 ounces
D) 14 ounces

28) Sheet copper, when used as a lining for flush tanks, shall weigh how much per square foot?

A) 8 ounces
B) 10 ounces
C) 12 ounces
D) 14 ounces

29) Brass floor flanges for water closets and similar fixtures must have a minimum thickness of:

A) ⅛ inch
B) ¼ inch
C) ½ inch
D) ¾ inch

30) Plastic floor flanges for water closets and similar fixtures must have a minimum thickness of:

A) ⅛ inch
B) ¼ inch
C) ½ inch
D) ¾ inch

31) Cast-iron floor flanges for water closets and similar fixtures must have a minimum caulking depth of:

A) ½ inch
B) 1 inch
C) 1½ inches
D) 2 inches

32) Screws and bolts used to secure flanges to a building structure must be:

A) 2 inches long
B) Concealed
C) Corrosion resistant
D) All of the above

33) Cleanout plugs may be made out of which of the following materials?

A) Lead
B) Brass
C) Copper
D) All of the above

34) Cleanout plugs may be made out of which of the following materials?

A) Lead
B) Copper
C) Plastic
D) All of the above

35) When a raised-head type of cleanout plug may create a tripping hazard, what type of head may be used in its place?

A) A countersunk head
B) A flush head
C) Either A or B
D) Neither A nor B

36) A flush pipe or flush fitting may be made out of what type of material?

A) Ferrous
B) Nonferrous
C) Either A or B
D) Neither A nor B

37) What type of surface must plumbing fixtures have?

A) White surfaces
B) Stainless-steel surfaces
C) Smooth, impervious surfaces
D) Porous surfaces

38) What type of surface must plumbing fixtures not have?

A) Stainless-steel surfaces
B) Smooth, impervious surfaces
C) Concealed fouling surfaces
D) China surfaces

39) Soldering bushings shall be made of what type of material?

A) Copper
B) Yellow brass
C) Red brass
D) Any of the above

40) Caulking ferrules shall be made of what type of material?

A) Wrought copper
B) Bronze
C) Red brass
D) Any of the above

41) A water service pipe may not be installed in ground that is:

A) Contaminated with solvents
B) Contaminated with fuels
C) Corrosive
D) All of the above

42) At what temperature is the pressure rating of a water service pipe measured?

A) 72 degrees Fahrenheit
B) 73.4 degrees Fahrenheit
C) 76.3 degrees Fahrenheit
D) 80 degrees Fahrenheit

43) Where water pressure exceeds 160 psi, piping material for a water service must be:

A) Copper
B) Polybutylene
C) Rated with a minimum working pressure equal to the highest available pressure
D) Any of the above

44) At what temperature is the pressure rating for hot water distribution pipes measured?

A) 110 degrees Fahrenheit
B) 120 degrees Fahrenheit
C) 140 degrees Fahrenheit
D) 180 degrees Fahrenheit

45) Valves used with plumbing pipes must be:

A) Of an approved type
B) Compatible with the type of piping used
C) Both A and B
D) Either A or B

Congratulations, you have just completed the multiple-choice exam on approved materials. Before you turn to the end of the chapter to check your answers, take the true-false test for your zone.

True-false test for zone 1

1) If copper tubing is used for underground drainage piping, it must have a weight of not less than DWV copper.

True False

2) Copper tubing used for above-ground drainage must have a weight of not less than Type-M copper.

True False

3) Copper tubing used for water piping must have a weight of not less than Type K copper.

True False

4) All hard-drawn copper tubing must be marked by means of a continuous and indelibly colored stripe at least one quarter of an inch wide.

True False

5) Type L copper is marked with a blue line.

True False

6) Type K copper is marked with a green line.

True False

7) Type M copper is marked with a red line.

True False

8) Caulking ferrules must be manufactured from either bronze or copper.

True False

9) Soldering bushings must be made of bronze or copper.

True False

10) Closet flanges must have a diameter of approximately 9 inches.

True False

11) Closet screws and bolts must be made of corrosion resistant materials.

True False

12) Cleanouts must be designed to be gas and watertight without the use of any gasket, packing, or washer.

True False

13) The bodies of backwater valves must be made of brass.

True False

14) Gate valves used on drainage piping must be of a full-way design and of a corrosive-resistant material.

True False

15) Fittings used for drainage must be designed to allow a drainage pitch of at least ⅛ of an inch per each foot of fall.

True False

True-false test for zone 2

1) Sheet lead used for safe pans must be coated with an asphalt paint or equivalent.

True False

2) Flashings formed out of sheet lead to seal vent terminals must have a weight of 3 pounds per square foot.

True False

3) Caulking ferrules may not be made of bronze or wrought copper.

True False

4) Nonmetallic shower pan material is not required to be permanently marked by the manufacturer.

True False

5) Non-plasticized chlorinated polyethylene sheets for shower pans shall have a minimum thickness of 0.040 inch.

True False

6) The commingling of different materials, without the use of approved adapters is prohibited.

True False

7) Stainless steel water tube is for use below-ground only.

True False

8) Air admittance valves must be accessible for inspection.

True False

9) Where permitted, soldering bushings must be made of yellow brass.

True False

10) Cleanout plugs used with borosilicate glass systems must be made of borosilicate glass.

True False

True-false test for zone 3

1) The regulations that pertain to approved materials extend only to materials used in the installation or alteration of plumbing systems and do not pertain to plumbing repairs.

True False

2) Where detrimental conditions exist for the installation of a water service pipe, approved alternative materials or routing must be used.

True False

3) Chlorinated polyvinyl chloride (CPVC) pipe is an approved material for water service use.

True False

4) Existing building sewers and drains may not connect to new building sewers and drains unless the existing materials conform to the new system in quality of material.

True False

5) Materials approved for use with above-ground drainage and vent piping may also be used for storm drainage conductors inside buildings.

True False

6) Building storm sewer piping may be made of Acrylonitrile Butadiene Styrene (ABS) plastic pipe.
True False

7) Type M copper may not be used for a building storm drain pipe.
True False

8) Closet screws and bolts must be made of brass.
True False

9) Special-use fixtures may not be made of soapstone.
True False

10) Floor flanges for toilets must be secured by no less than six screws or bolts.
True False

You have just completed all of the tests on the subject of approved materials. Please go to the end of the chapter and review the correct answers to the questions. Compare these answers with the ones you gave, and mark any incorrect answers on your test.

Answers to multiple-choice exam

1) E
2) C
3) C
4) C
5) D
6) D
7) A
8) D
9) D
10) B
11) A
12) F
13) F
14) C
15) C
16) C
17) C
18) D
19) D
20) A
21) B
22) B
23) B
24) C
25) B
26) C
27) A
28) B
29) A
30) B
31) D
32) C
33) B
34) C
35) A
36) B
37) C
38) C
39) C
40) C
41) D
42) B
43) C
44) D
45) B

Answers to true-false test for zone 1

1) T	6) F	11) T
2) F	7) T	12) T
3) F	8) T	13) F
4) T	9) T	14) T
5) T	10) F	15) F

Answers to true-false test for zone 2

1) T	5) T	8) T
2) T	6) T	9) F
3) F	7) F	10) T
4) F		

Answers to true-false test for zone 3

1) F	5) T	8) T
2) T	6) T	9) F
3) T	7) F	10) F
4) T		

Table 5-1

Materials Approved for Above-Ground Vents in Zone 1

Cast-iron materials
ABS materials*
PVC materials*
Copper materials
Galvanized materials
Lead materials
Brass materials

*Note—these materials may not be used with buildings having more than three floors above grade.

Table 5-2

Materials Approved for Underground Vents in Zone 1

Cast-iron materials
ABS materials
PVC materials
Copper materials
Aluminum materials
Borosilicate-glass materials

Table 5-3

Materials Approved for Above-Ground Vents in Zone 2

Cast-iron materials
ABS materials
PVC materials
Copper materials
Galvanized materials
Lead materials
Aluminum materials
Brass materials

Materials Approved for Underground Vents in Zone 2

Table 5-4

Cast-iron materials

ABS materials*

PVC materials*

Copper materials

Brass materials

Lead materials

*Note—these materials may not be used with buildings having more than three floors above grade.

Materials Approved for Above-Ground Vents in Zone 3

Table 5-5

Cast-iron materials

ABS materials

PVC materials

Copper materials

Galvanized materials

Lead materials

Aluminum materials

Borosilicate-glass materials

Brass materials

Materials Approved for Underground Vents in Zone 3

Table 5-6

Cast-iron materials

ABS materials

PVC materials

Copper materials

CHAPTER 6

Approved joints and connections

TO install a plumbing system that is in compliance with the plumbing code, you must understand the use of approved joints and connections. A lot of plumbers are accustomed to working with solvent-weld joints and solder joints, but not many plumbers know how to use the other types of joints discussed in the plumbing code. Of course, all plumbers with field experience know how to make a threaded joint, and they know that pipe dope or sealant tape is always applied to male threads. But do you know that it is a violation of the plumbing code to apply pipe dope to female threads? Did you know this? Got you thinking, didn't I?

In the old days fittings consisted mostly of caulked lead joints and threaded joints. Nowadays, almost anything goes. There are rubber couplings, solvent-weld joints, rubber doughnuts, and a lot of other possibilities for making legal, watertight joints. It is up to you, the plumber, to know what joints to use for various field conditions.

Let me tell you a story about two plumbers in Maine who were not too familiar with polybutylene (PB) pipe. These plumbers worked under my supervision for a couple of years. They were both good plumbers, but neither one of them had extensive field experience. I think they had been in the trade about three or four years. This is plenty of time to gather general plumbing ability, but a few years in the field is far from enough to prepare a plumber for all the situations that might be encountered. To make matters worse, these plumbers had worked all of their plumbing career in Maine.

Maine is not one of the more progressive, advanced states in which to work. The slow pace tends to keep technology several years behind many other states in terms of new plumbing materials and methods. I've lived in Maine for about eight years, but most of my experience was earned in Virginia, a state that tends to keep up with the times a bit better than Maine.

I started working with PB pipe a long time ago in Virginia. When PB pipe was first introduced, the methods used to make joints left something to be desired. There were reports that the joints wouldn't endure the test of time. In other words, plumbers were saying that the fittings would blow off the pipe and flood buildings. I never had this happen on any of my jobs, but I assume the reports were based on truth. To this day, I wonder if the problems came from people improperly trained to make the joints. At any rate, PB pipe is no stranger to me; I think it's great.

To get back to my story, these two plumbers had never worked with PB pipe and fittings. Maine being a traditional state, there isn't much of this material installed in homes. There is, however, a good deal of PB pipe installed in mobile homes and some modular homes. This is what brought the two plumbers into contact with what they called a strange, gray, plastic pipe. They didn't even know what PB pipe was!

One of the plumbers responded to a call at a mobile home. The job was to install a silcock. When the plumber saw the gray, plastic pipe, he assumed it could be worked with like other plastic tubing, so he made the connection with a stainless steel clamp and an insert fitting. After finishing his work, the plumber cut the water back on, tested his work, and left.

The next day, the second plumber was dispatched to the property to fix a leak in the new work done by the other plumber. He found that the leak was coming from the insert fitting. Thinking that the clamp wasn't tight enough, he tried to tighten it. He thought the leak was fixed and left.

The following day, the owner of the property called and complained that the pipe was still leaking. Since this was the second call-back, I responded to the call personally. When I arrived, the problem was obvious to me. PB pipe is stiff and a hose clamp is rarely a sufficient way to mate the pipe to an insert fitting. I took the joint apart and replaced the clamp with a PB crimp ring. Once the ring was crimped into place, the leak was history.

If either of the plumbers in this story had known the proper procedure for joining PB pipe with an insert fitting, the problem would not have occurred or escalated. Their lack of knowledge made the company look bad and cost the company money. This is not something employers want to have happen.

When you venture out as a plumber, you must have well-rounded knowledge. Most of this knowledge comes from field experience, but you can help it along by becoming familiar with the types of materials and work you will be doing. Part of this self-study involves joints and connections, since they are, after all, a big part of a plumber's job. Now, let's turn our attention to the multiple-choice exam on joints and connections.

Multiple-choice exam

1) According to the plumbing code, all joints must be of which of the following types:

A) ASP certified type
B) Approved type
C) Metallic type
D) Classified type

2) When we talk of a joint being tight, which of the following apply:

A) Gas tight
B) Water tight
C) Both A and B
D) None of the above

3) When preparing the end of a pipe to make a joint, the end of the pipe should be _____. Fill in the blank.

A) Cut square
B) Reamed
C) Chamfered
D) All of the above

4) Pipe used to make a joint should not be _____. Fill in the blank.

A) Plastic
B) Overcut
C) Undercut
D) Labeled

5) When pipe is installed in a flood-hazard zone, it must be capable of resisting which of the following:

A) Extreme heat
B) Hydrostatic loads
C) Hydrodynamic loads
D) Both B and C

6) When pipe is installed in a flood-hazard zone, it must be capable of resisting which of the following:

A) Extreme heat
B) Stress
C) Cold water
D) Both A and C

7) A compression coupling is a type of _____. Fill in the blank.

A) Threaded joint
B) Mechanical joint
C) Illegal joint
D) Screw joint

8) Elastomeric seals are used with which of the following types of joints:

A) Brazed
B) Welded
C) Caulked
D) Mechanical

9) If plastic pipe is to be threaded, it must carry a rating of at least _____. Fill in the blank.

A) Schedule 20
B) Schedule 40
C) Schedule 60
D) Schedule 80

10) When a solvent-weld joint is to be made, the joint surfaces must be:

A) Clean
B) Free of moisture
C) Roughed up
D) Both A and B

11) Solvent-cement joints are not permitted:

A) Above ground
B) Below ground
C) In attics
D) None of the above

12) Solvent-cement joints must be made while:

A) A master plumber is present
B) A journeyman plumber is present
C) The cement is wet
D) Both A and C

13) When making a threaded joint, the pipe dope or thread sealant should be applied to which of the following:

A) The male threads
B) The female threads
C) Both A and B
D) Either A or B

14) If a joint is made between asbestos-cement pipe and fittings, it must be made with which of the following:

A) A rubber coupling
B) A sleeve coupling
C) An approved coupling
D) Both B and C

15) Couplings used to make joints with asbestos-cement pipe must be:

A) Of an approved type
B) Of the same composition as the pipe
C) Both A and B
D) None of the above

16) Before a brazed joint is made, all joint surfaces must be:

A) Inspected
B) Cleaned
C) Dampened
D) All of the above

17) Mechanical joints must be installed in accordance with:

A) Local code requirements
B) The manufacturer's instructions
C) Both A and B
D) None of the above

18) When a joint is welded, the filler material must be:

A) Silver
B) Lead
C) Brass
D) Approved

19) Joints for hub and spigot cast-iron pipe must be packed firmly with which of the following:

A) Oakum
B) Hemp
C) Both A and B
D) Either A or B

20) When molten lead is used to make a joint with cast-iron pipe, the lead must be poured in a single operation to a depth of at least:

A) ½ inch
B) 1 inch
C) 1½ inches
D) 2 inches

21) Joints made between concrete pipe and fittings must be made with which of the following types of seals:

A) Electromagnetic
B) Elastomeric
C) Elam
D) None of the above

22) Before a soldered joint is made, which of the following must be applied to the joint surfaces:

A) Flux
B) Pipe dope
C) Primer
D) Both A and C

23) The joining of water pipe with soldered joints must be done with a solder that is rated as lead-free. What is the maximum percentage of lead that the solder may contain?

A) 0.1 percent
B) 0.2 percent
C) 1 percent
D) 2 percent

24) Before making a joint with CPVC pipe and fittings, all joint surfaces must be:

A) Clean
B) Free of moisture
C) Primed
D) All of the above

25) Solvent-cement joints used with CPVC pipe are not permitted in which of the following locations:

A) Above ground
B) Below ground
C) In attics
D) None of the above

26) If lead pipe is being joined, the thickness of the joint must be at least _____. Fill in the blank.

A) 3 inches long
B) As thick as the lead being joined
C) Smoothed to remove any rough edges
D) Both A and C

27) When a lead joint is burned, the filler material must be:

A) Silver solder
B) The same material as the pipe
C) Heated to 750 degrees Fahrenheit
D) Heated to 1000 degrees Fahrenheit

28) When heat-fusion joints are made with PB pipe, they must be of which of the following types:

A) Socket-fusion
B) Butt-fusion
C) Both A and B
D) Either A or B

29) When a heat-fusion joint is made with polyethylene (PE) pipe, which of the following is necessary:

A) Clean surface areas to be joined and make sure they are free of moisture
B) Heat joint surfaces to melt temperature until they are joined
C) Leave joint undisturbed until it is cool
D) All of the above

30) Except when joined by heat fusion, the ends of PE pipe must be _____ before a joint is made. Fill in the blank.

A) Roughed up with a light-grit sandpaper
B) Heated
C) Chamfered
D) All of the above

31) PE pipe may not be bent within a certain number of pipe diameters from a fitting, what is that number?

A) 2 pipe sizes
B) 8 pipe sizes
C) 10 pipe sizes
D) 12 pipe sizes

32) If you are installing PE pipe that has a diameter of 1 inch and it is necessary to bend the pipe, how far from the nearest fitting can the bend in the pipe be placed?

A) 4 inches
B) 6 inches
C) 10 inches
D) 12 inches

33) When a joint is made with PVC pipe, what must be applied to the joint surface prior to making the joint?

A) Flux
B) Clear primer
C) Purple primer
D) Plain cleaner

34) What color of solvent cement is prohibited from use with PVC pipe?

A) Clear
B) Gold
C) Purple
D) Green

35) What type of joint is required to join two pipes of different materials?

A) Solvent-weld
B) Screw
C) Butt
D) Mechanical

36) Connectors or adapters used to join two pipes of different materials must have which of the following:

A) A stainless steel clamp
B) An elastomeric seal
C) A rubber insulator
D) All of the above

37) What type of ferrule is required when copper tubing is joined with cast iron?

A) Nylon
B) Brass
C) Copper
D) Any of the above

38) When copper tubing is joined with cast iron, through the use of a ferrule, which of the following connection methods is acceptable?

A) A mechanical joint
B) A caulked joint
C) A soldered joint
D) Either A or B

39) When copper tubing is joined with galvanized steel pipe, which of the following is required:

A) A brass converter fitting
B) A dielectric fitting
C) A brazed joint
D) Either A or B

40) Which of the following is an approved means for joining cast-iron pipe with galvanized steel pipe?

A) Caulked joints
B) Threaded joints
C) Mechanical joints
D) All of the above

41) When plastic pipe is joined with cast-iron pipe, which of the following types of joints is acceptable:

A) Mechanical joints
B) Caulked joints
C) Solvent-weld joints
D) Either A or B

42) When lead pipe is joined with another type of pipe, which of the following types of joints is acceptable:

A) Solvent-weld joints
B) Mechanical joints
C) Threaded joints
D) Welded joints

43) When lead pipe is joined with another type of pipe, which of the following types of joints is acceptable:

A) Wiped joints to a caulking ferrule
B) Soldering nipple
C) Bushing
D) Any of the above

44) Which of the following types of joints are prohibited?

A) Mechanical joints
B) Wiped joints
C) Cement joints
D) Soldered joints

45) Which of the following types of joints are prohibited?

A) Solvent-cement joints between different types of plastic pipe
B) Mastic or hot-pour bituminous joints
C) Concrete joints
D) All of the above

46) Connections between drain and floor outlet plumbing fixtures must be made with which of the following:

A) Mechanical joints
B) Floor flanges
C) Schedule-40 joints
D) Any of the above

47) Connections between the drain and wall-hung water closets must be made with which of the following:

A) Extension nipples
B) Horn adapters
C) Wax rings
D) Either A or B

48) Slip joints may be installed in which of the following locations:

A) On the trap outlet
B) On the trap inlet
C) Within the trap seal
D) Any of the above

49) Flexible couplings must be resistant to which of the following:

A) Water
B) Sewage
C) Oil
D) Shrinkage

50) Cement mortar joints are prohibited, unless you are:

A) Repairing an existing joint of the same type
B) Connecting to an existing line constructed with cement joints
C) Either A or B
D) None of the above

51) When plastic pipe is used as a water service, which of the following types of fittings may be used?

A) Insert fittings
B) Compression fittings
C) Either A or B
D) None of the above

52) Polyethylene water services that are fitted with stiffeners must be installed so that the stiffeners extend:

A) 1½ inches past the clamp or nut
B) No further than the clamp or nut
C) To a point no more than one pipe size past the clamp or nut
D) None of the above

53) Which of the following types of joints are acceptable for use with cast-iron pipe?

A) Caulked joints
B) Elastomeric compression joints
C) Neoprene gasket and stainless steel retaining sleeve joints
D) All of the above

54) Expansion joints must be:

A) Soldered
B) Accessible
C) Recorded
D) Avoided

This concludes the multiple-choice test on joints and connections. Please move to the true-false test that applies to your region and answer those questions before you check your answers.

True-false test for zone 1

1) When making a solder or sweat joint, it is required that all joint surfaces be cleaned bright. This cleaning may be done by either manual or mechanical means.

   True False

2) The flux used to make solder or sweat joints must be of a noncorrosive type.

   True False

3) The flux used to make solder or sweat joints must have a lead content that does not exceed .3 of 1 percent.

   True False

4) When working with a difficult joint, it is permissible to add a leak-sealing additive to the joint packing.

   True False

5) Female PVC screwed fittings are prohibited for drainage use.

   True False

6) Female PVC screwed fittings are prohibited for use with water distribution pipes.

   True False

7) Unions are not allowed for use with drainage piping.

   True False

8) When installing water piping, the use of slip joints is limited to only exposed fixture supplies.

   True False

9) A fitting with an enlargement, chamber, or recess with a ledge, shoulder, or reduction of pipe area, that creates an obstruction in the flow of drainage may not be installed.

True False

10) Only brass or cast-iron body cleanouts may be used as reducers or adapters when going from cast-iron pipe to plastic pipe in a drainage system.

True False

True-false test for zone 2

1) Unions used in drainage systems must have metal-to-metal seats.

True False

2) Unions used in water distribution systems must be metal-to-metal with ground seats.

True False

3) Drilling, tapping, or installing saddle fittings in the water distribution system is prohibited.

True False

4) Only horn adapters shall be used to make connections between drainage piping and wall-hung water closets.

True False

5) End-cap adapters and coupling assemblies must be used when making joints with aluminum pipe.

True False

6) When making a connection between a fixture and a drainage system, any of the following methods of creating a gastight seal are acceptable: approved gaskets, approved washers, approved setting compounds, plaster, and commercial putty.

True False

7) Transition fittings for dissimilar pipe in building sewers are prohibited from use with PVC pipe.

True False

8) If polyethylene pipe is being used as a water service, there are requirements that control the amount of bending that may be done with the pipe. When bending PE pipe under these conditions, the pipe curvature cannot be less than 30 pipe diameters, or the coil radius when bending with the coil.

True False

9) If polyethylene pipe is being used as a water service, there are requirements that control the amount of bending that may be done with the pipe. When bending PE pipe under these conditions, no bends shall be made within 8 pipe diameters of any fitting or valve.

True False

10) When working with borosilicate glass joints, the joints between glass to glass must be made with a bolted compression-type stainless steel coupling with contoured acid-resistant elastomer compression ring and a fluorocarbon polymer inner seal ring.

True False

True-false test for zone 3

1) Joints made between bituminized fiber pipe must be made with flared-type couplings of the same material.

True False

2) A joint between two pipes that allows one pipe to move without moving the other pipe is called an expansion joint.

True False

3) When caulking a lead joint with cast-iron pipe that has a 4-inch diameter, the depth of the lead in the joint should be 2¼ inches.

True False

4) For a material to be considered lead-free, it must contain no more than .2 of 1 percent of lead.

True False

5) When making a wiped joint with lead pipe, the joint must be wiped fully with an exposed surface on each side of the joint not less than ¾ inch.

True False

6) When joining two pipes of dissimilar materials, the connection must be made with a mechanical joint.

True False

7) Cement or concrete joints are prohibited.

True False

8) Solvent-cement joints between different types of plastic pipe can be made with the use of a transitional cement.

True False

9) Joints made between different diameter pipes may be made with elastomeric rolling O-rings.

True False

10) Mastic or hot-pour bituminous joints may be used when the ambient air temperature is above 32 degrees Fahrenheit.

True False

This brings us to the end of the tests on joints and connections. It is now time to compare your answers with the correct answers at the end of the chapter. Check your work carefully and take the time now to find out why you answered any questions incorrectly.

The joints and connections section of the code book is not extensive, but the information is important to routine plumbing work. While many of the joints and connections covered in this section are rarely used by the average plumber, there is a good likelihood that questions that pertain to them will show up on the exam.

The next chapter is going to test your knowledge about installing sanitary drainage systems. All of the code is important to learn and understand for the licensing exam, but the next chapter is one of the most important. Since so much of plumbing involves sanitary drainage, vents, and water distribution, knowledge of the information in the next chapter will play a key role in whether or not you pass the licensing exam. After you check your answers for this chapter, turn to chapter 7 and thoroughly absorb the material.

Answers to multiple-choice exam

1) B	15) C	29) D	42) B
2) C	16) B	30) C	43) D
3) D	17) C	31) C	44) C
4) C	18) D	32) C	45) D
5) D	19) D	33) C	46) B
6) B	20) B	34) C	47) D
7) B	21) B	35) D	48) D
8) D	22) A	36) B	49) C
9) D	23) B	37) B	50) C
10) D	24) D	38) D	51) C
11) D	25) D	39) D	52) B
12) C	26) B	40) D	53) D
13) A	27) B	41) D	54) B
14) D	28) D		

Answers to true-false test for zone 1

1) T	5) F	8) T
2) T	6) T	9) T
3) F	7) T	10) F
4) F		

Answers to true-false test for zone 2

1) T	5) T	8) T
2) T	6) F	9) F
3) T	7) F	10) T
4) F		

Answers to true-false test for zone 3

1) F	5) T	8) F
2) F	6) T	9) F
3) T	7) T	10) F
4) T		

CHAPTER 7

Installing sanitary drainage systems

THIS chapter is going to be a bit more intense than the first few chapters of this book. The reason for this is simple, you are going to see a lot of questions on the real exam that deal with drainage systems, so it is best to be hit with them here and now, in order to be well prepared on the day of the test.

When it comes to installing sanitary drainage systems, there is a lot to learn. For example, some types of fittings can be used to turn a pipe from a horizontal run to a vertical run, but not from a horizontal to a horizontal run. The fact is, the rules and regulations that pertain to sanitary drainage systems are extensive.

Before you can consider yourself competent to take full charge of a drainage system, you must know how to size drainage pipes and traps. This work is not difficult once you understand the basic principles, but it can be confusing to someone who is not accustomed to doing it. Refer to Tables 7-1 through 7-15 at the end of this chapter for specific information by zone.

If you casually read the drainage chapter in your code book, you might feel that the chapter is short, concise, and easy to understand. This can be true if you have the foundation of knowledge needed to use the information provided in the code. But if you are not fluent in fixture units and the types of fittings allowed for changes in direction, your feelings for this chapter might change quickly.

The charts and tables provided that identify fixture-unit ratings don't take up much space in the code book. Each table is filled out with a bunch of numbers. Have you studied these numbers enough to know how many fixture units a particular type of drain can carry? You'd better, if you want a high score on the exam.

Don't be fooled by the brevity of the drainage systems chapter in your code book. While the words may be few, the rules and regulations are many. You will also be wise to learn the chapter thoroughly, since much of your licensing exam is likely to be based on drainage systems.

I spoke earlier in the book about the three major plumbing codes and how there are some differences between each of them. Of course, on the local level the plumbing code can be even more different from any of the three primary codes. The multiple-choice exams given throughout this book encompass most of the common ground shared by the primary plumbing codes. However, there might be some questions that don't pertain to your particular region. When this is the case, you will find questions in the true-false tests that modify the test to suit your regional needs. I bring this up now as a reminder, only so that you will not be confused if there is a question or two that you cannot find the answer to in your code book.

I also talked earlier about how plumbing exams often manipulate words to confuse test takers. To simulate what I believe you might encounter on the real exam, I have used these same tactics in the mock exams you are about to take. Pay attention, take your time, and see how you do.

Multiple-choice exam

1) When a drainage system is installed to convey chemical waste, the system must be:
   - A) Installed with Schedule-40 plastic pipe
   - B) Tested at a pressure of 20 PSI
   - C) Completely separated from the sanitary drainage system
   - D) All of the above

2) When a drainage fixture-unit value is assigned for a drain that is to receive the waste from a continuous or semicontinuous flow, such as from a pump, the fixture-unit rating must be based on the assumption that one-fixture is equivalent to how many gallons of flow per minute?

A) 2.5 gallons
B) 3.5 gallons
C) 5 gallons
D) 7.5 gallons

3) When pipe sizing for a sanitary drainage system is computed, the primary factor that determines the appropriate pipe size is:

A) The fixture-unit load
B) The slope of the drain
C) The type of pipe being used for the drain
D) All of the above

4) Which of the following must be added into the fixture-unit load when sizing drainage pipe?

A) Potential future fixtures
B) Future fixtures that are on the blueprints
C) Future fixtures that are roughed in
D) All of the above

5) For the purposes of sizing a drainage stack, horizontal offsets must be classified as:

A) Fixture loads
B) Piping obstructions
C) Future fixture loads
D) Part of the stack

6) A plumbing fixture that has a trap with a diameter of 1¼ inches or less is rated at how many fixture units?

A) 1
B) 2
C) 3
D) 4

7) A plumbing fixture that has a trap with a diameter of 1½ inches is rated at how many fixture units?

A) 1
B) 2
C) 3
D) 4

8) A plumbing fixture that has a trap with a diameter of 2 inches is rated at how many fixture units?

A) 1
B) 2
C) 3
D) 4

9) A plumbing fixture that has a trap with a diameter of 2½ inches is rated at how many fixture units?

A) 1
B) 2
C) 3
D) 4

10) A plumbing fixture that has a trap with a diameter of 3 inches is rated at how many fixture units?

A) 4
B) 5
C) 6
D) 7

11) A plumbing fixture that has a trap with a diameter of 4 inches is rated at how many fixture units?

A) 4
B) 5
C) 6
D) 7

12) Which of the following numbers indicates the minimum slope, per inch per foot, required for a pipe with a diameter of 2½ inches or less?

A) 1/16 of an inch per foot
B) ⅛ of an inch per foot
C) ¼ of an inch per foot
D) ½ of an inch per foot

13) Which of the following numbers indicates the minimum slope, per inch per foot, required for a pipe with a diameter of between 3 and 6 inches?

A) 1/16 of an inch per foot
B) ⅛ of an inch per foot
C) ¼ of an inch per foot
D) ½ of an inch per foot

14) Which of the following numbers indicates the minimum slope, per inch per foot, required for a pipe with a diameter of 8 inches or more?

A) 1/16 of an inch per foot
B) ⅛ of an inch per foot
C) ¼ of an inch per foot
D) ½ of an inch per foot

15) What is the drainage fixture-unit value of a commercial automatic clothes washer standpipe?

A) 1 fixture unit
B) 2 fixture units
C) 3 fixture units
D) 4 fixture units

16) What is the drainage fixture-unit value of a domestic automatic clothes washer standpipe?

A) 1 fixture unit
B) 2 fixture units
C) 3 fixture units
D) 4 fixture units

17) What is the drainage fixture-unit value of a bathroom group?

A) 3 fixture units
B) 4 fixture units
C) 5 fixture units
D) 6 fixture units

18) What is the drainage fixture-unit value of a domestic bathtub?

A) 1 fixture unit
B) 2 fixture units
C) 3 fixture units
D) 4 fixture units

19) What is the drainage fixture-unit value of a domestic bidet?

A) 1 fixture unit
B) 2 fixture units
C) 3 fixture units
D) 4 fixture units

20) What is the drainage fixture-unit value of a combination sink and tray?

A) 1 fixture unit
B) 2 fixture units
C) 3 fixture units
D) 4 fixture units

21) What is the drainage fixture-unit value of a dental unit?

A) 1 fixture unit
B) 2 fixture units
C) 3 fixture units
D) 4 fixture units

22) What is the drainage fixture-unit value of a dishwasher?

A) 1 fixture unit
B) 2 fixture units
C) 3 fixture units
D) 4 fixture units

23) What is the drainage fixture-unit value of a drinking fountain?

A) ⅓ fixture unit
B) ½ fixture unit
C) 1 fixture unit
D) 1½ fixture units

24) What is the drainage fixture-unit value of a floor drain?
A) 1 fixture unit
B) 2 fixture units
C) 3 fixture units
D) 4 fixture units

25) What is the drainage fixture-unit value of a kitchen sink?
A) 1 fixture unit
B) 2 fixture units
C) 3 fixture units
D) 4 fixture units

26) What is the drainage fixture-unit value of a laundry tray?
A) 1 fixture unit
B) 2 fixture units
C) 3 fixture units
D) 4 fixture units

27) What is the drainage fixture-unit value of a lavatory?
A) 1 fixture unit
B) 2 fixture units
C) 3 fixture units
D) 4 fixture units

28) What is the drainage fixture-unit value of a mop basin?
A) 1 fixture unit
B) 2 fixture units
C) 3 fixture units
D) 4 fixture units

29) What is the drainage fixture-unit value of a service sink?
A) 1 fixture unit
B) 2 fixture units
C) 3 fixture units
D) 4 fixture units

30) What is the drainage fixture-unit value of a single shower head?
A) 1 fixture unit
B) 2 fixture units
C) 3 fixture units
D) 4 fixture units

31) What is the drainage fixture-unit value of a sink?
A) 1 fixture unit
B) 2 fixture units
C) 3 fixture units
D) 4 fixture units

32) What is the drainage fixture-unit value of a urinal?
A) 1 fixture unit
B) 2 fixture units
C) 3 fixture units
D) 4 fixture units

33) What is the drainage fixture-unit value of a private water closet?

A) 1 fixture unit
B) 2 fixture units
C) 3 fixture units
D) 4 fixture units

34) What is the drainage fixture-unit value of a public water closet?

A) 3 fixture units
B) 4 fixture units
C) 5 fixture units
D) 6 fixture units

35) What is the minimum trap size required for a commercial automatic clothes washer?

A) 1¼ inch
B) 1½ inch
C) 2 inch
D) 3 inch

36) What is the minimum trap size required for a domestic automatic clothes washer?

A) 1¼ inch
B) 1½ inch
C) 2 inch
D) 3 inch

37) What is the minimum trap size required for a bathtub?

A) 1¼ inch
B) 1½ inch
C) 2 inch
D) 3 inch

38) What is the minimum trap size required for a bidet?

A) 1¼ inch
B) 1½ inch
C) 2 inch
D) 3 inch

39) What is the minimum trap size required for a combination sink and tray?

A) 1¼ inch
B) 1½ inch
C) 2 inch
D) 3 inch

40) What is the minimum trap size required for a dental unit?

A) 1¼ inch
B) 1½ inch
C) 2 inch
D) 3 inch

41) What is the minimum trap size required for a dishwasher?

A) 1¼ inch
B) 1½ inch
C) 2 inch
D) 3 inch

42) What is the minimum trap size required for a drinking fountain?

A) 1¼ inch
B) 1½ inch
C) 2 inch
D) 3 inch

43) What is the minimum trap size required for an emergency floor drain?

A) 1¼ inch
B) 1½ inch
C) 2 inch
D) 3 inch

44) What is the minimum trap size required for a floor drain?

A) 1¼ inch
B) 1½ inch
C) 2 inch
D) 3 inch

45) What is the minimum trap size required for a kitchen sink?

A) 1¼ inch
B) 1½ inch
C) 2 inch
D) 3 inch

46) What is the minimum trap size required for a laundry tray?

A) 1¼ inch
B) 1½ inch
C) 2 inch
D) 3 inch

47) What is the minimum trap size required for a lavatory?

A) 1¼ inch
B) 1½ inch
C) 2 inch
D) 3 inch

48) What is the minimum trap size required for a mop basin?

A) 1¼ inch
B) 1½ inch
C) 2 inch
D) 3 inch

49) What is the minimum trap size required for a service sink?

A) 1¼ inch
B) 1½ inch
C) 2 inch
D) 3 inch

50) What is the minimum trap size required for a single shower head?

A) 1¼ inch
B) 1½ inch
C) 2 inch
D) 3 inch

51) What is the minimum trap size required for a sink?

A) 1¼ inch
B) 1½ inch
C) 2 inch
D) 3 inch

52) What is the minimum trap size required for a urinal?

A) 1¼ inch
B) 1½ inch
C) 2 inch
D) 3 inch

53) A drain pipe that has a diameter of 1¼ inches and a fall of ¼ inch per foot can carry how many fixture units?

A) 1
B) 1½
C) 2
D) 3

54) A drain pipe that has a diameter of 1¼ inches and a fall of ½ inch per foot can carry how many fixture units?

A) 1
B) 1½
C) 2
D) 3

55) A drain pipe that has a diameter of 1½ inches and a fall of ¼ inch per foot can carry how many fixture units?

A) 1
B) 1½
C) 2
D) 3

56) A drain pipe that has a diameter of 1½ inches and a fall of ½ inch per foot can carry how many fixture units?

A) 1
B) 1½
C) 2
D) 3

57) A drain pipe that has a diameter of 2 inches and a fall of ¼ inch per foot can carry how many fixture units?

A) 10
B) 20
C) 21
D) 26

58) A drain pipe that has a diameter of 2 inches and a fall of ½ inch per foot can carry how many fixture units?

A) 10
B) 20
C) 21
D) 26

59) A drain pipe that has a diameter of 3 inches and a fall of ⅛ inch per foot can carry how many fixture units?

A) 26
B) 36
C) 42
D) 50

60) A drain pipe that has a diameter of 3 inches and a fall of ¼ inch per foot can carry how many fixture units?

A) 26
B) 36
C) 42
D) 50

61) A drain pipe that has a diameter of 3 inches and a fall of ½ inch per foot can carry how many fixture units?

A) 26
B) 36
C) 42
D) 50

62) A drain pipe that has a diameter of 4 inches and a fall of ⅛ inch per foot can carry how many fixture units?

A) 160
B) 180
C) 216
D) 250

63) A drain pipe that has a diameter of 4 inches and a fall of ¼ inch per foot can carry how many fixture units?

A) 160
B) 180
C) 216
D) 250

64) A drain pipe that has a diameter of 4 inches and a fall of ½ inch per foot can carry how many fixture units?

A) 216
B) 250
C) 276
D) 295

65) Which of the following fittings can be used for a change in direction in drainage piping from horizontal to vertical?

A) 16th bend
B) 8th bend
C) 6th bend
D) All of the above

66) Which of the following fittings can be used for a change in direction in drainage piping from horizontal to vertical?

A) Quarter bend
B) Short sweep
C) Long sweep
D) All of the above

67) Which of the following fittings can be used for a change in direction in drainage piping from horizontal to vertical?

A) Quarter bend
B) Wye
C) Combination wye and 8th bend
D) All of the above

68) Which of the following fittings can be used for a change in direction in drainage piping from horizontal to vertical in only certain circumstances?

A) Quarter bend
B) Wye
C) Combination wye and 8th bend
D) Sanitary tee

69) Which of the following fittings may not be used for a change in direction in drainage piping from a vertical to horizontal direction?

A) Quarter bend
B) Wye
C) Combination wye and 8th bend
D) Sanitary tee

70) Which of the following fittings may be used for a change in direction in drainage piping from a vertical to horizontal direction?

A) 16th bend
B) Sanitary tee
C) Side-inlet quarter bend
D) All of the above

71) Which of the following fittings may be used for a change in direction in drainage piping from a vertical to horizontal direction?

A) 8th bend
B) 6th bend
C) Wye
D) All of the above

72) Which of the following fittings may be used for a change in direction in drainage piping from a vertical to horizontal direction?

A) 8th bend
B) Combination wye and 8th bend
C) Wye
D) All of the above

73) Which of the following fittings may be used for a change in direction in drainage piping from a vertical to horizontal direction only when the pipe is at least 3 inches in diameter?

A) Short sweep
B) Combination wye and 8th bend
C) Wye
D) Quarter bend

74) When a horizontal branch is connected to horizontal stack offsets and to the bases of stacks, the connection point must not be less than _____ pipe diameters downstream from the stack. Fill in the blank.

A) 4
B) 6
C) 8
D) 10

75) All drainage pipe installed for future fixtures must terminate with a:

A) Plug
B) Cap
C) Either A or B
D) None of the above

76) Which of the following shall not be considered as dead-ends in drainage piping?

A) Cleanout extensions
B) Approved future fixture rough-ins
C) Both A and B
D) Neither A nor B

77) All building subdrains must discharge into an approved:

A) Cesspool
B) Septic tank
C) Sump
D) Sewer

78) Sewage pumps and sewage ejectors must automatically discharge the contents of their sumps into:

A) A French drain
B) A holding tank
C) A septic tank
D) The building drainage system

79) What is the minimum discharge opening required for a grinder pump that receives the discharge from a water closet?

A) 1 inch
B) 1¼ inches
C) 1½ inches
D) 2 inches

80) What type of valve must be installed with a grinder pump or sewage ejector?

A) Air-volume control valve
B) Stop-and-waste valve
C) Check valve
D) Uniflow valve

81) What is the maximum interval of time allowed between the periods of evacuating a sump that serves a building subdrain?

A) 4 hours
B) 8 hours
C) 12 hours
D) 24 hours

82) A sewage pump or ejector that receives the discharge of a water closet must be capable of handling spherical solids with a diameter of up to and including _____ inches. Fill in the blanks.

A) 1¼
B) 1½
C) 2
D) 2½

83) A sewage pump or ejector that does not receive the discharge of a water closet must be capable of handling spherical solids with a diameter of up to and including _____ inches. Fill in the blanks.

A) 1
B) 1¼
C) 1½
D) 2

84) Which of the following fittings may be used for a change in direction in drainage piping from a horizontal to horizontal direction?

A) Sanitary tee
B) Combination wye and 8th bend

C) Both A and B
D) Neither A nor B

85) Which of the following fittings may be used for a change in direction in drainage piping from a horizontal to horizontal direction?

A) 16th bend
B) Combination wye and 8th bend
C) Both A and B
D) Neither A nor B

You've just completed the multiple-choice exam that pertains to sanitary drainage systems. How do you think you did? Well, before you check your answers, take the true-false quiz that relates to your region. Then proceed to the end of the chapter and check your answers.

True-false test for zone 1

1) It is a code violation to install galvanized wrought-iron or galvanized steel pipe below grade.

True False

2) Galvanized wrought-iron pipe and galvanized steel pipe must be installed at least 6 inches above grade.

True False

3) The use of ABS or PVC pipe is limited to structures that do not have more than two stories above ground.

True False

4) Vitrified clay pipe may only be used above ground.

True False

5) Vitrified clay pipe may not be pressurized by a pump or ejector.

True False

6) Fittings used on screwed pipe must be of the recessed drainage type.

True False

7) A fixture trap with a diameter of 1¼ inches may handle up to 2 fixture units.

True False

8) A fixture trap with a diameter of 1½ inches may handle up to 3 fixture units.

True False

9) A fixture trap with a diameter of 2 inches may handle up to 6 fixture units.

True False

10) A fixture trap with a diameter of 3 inches may handle up to 6 fixture units.

True False

11) A fixture trap with a diameter of 4 inches may handle up to 8 fixture units.

True False

12) A bidet has a fixture-unit rating of 2 and requires a 1½-inch trap.

True False

13) An automatic clothes washer requires a 2-inch trap.

True False

14) A mobile home park trap (for one trailer) must have a diameter of 4 inches and carry a fixture-unit rating of 6.

True False

15) The trap for a shower must have a minimum diameter of 2 inches.

True False

16) A commercial bar sink has a fixture-unit rating of 2.

True False

17) Private bar sinks have fixture-unit ratings of 1½.

True False

18) Clinic sinks require a trap with a minimum diameter of 2 inches.

True False

19) Fixture-unit ratings and trap sizes for urinals vary, depending on the style of urinal being installed.

True False

20) Sixty-degree branches may be used only when they are installed in a true vertical position.

True False

21) Two fixtures set back to back or side by side can share a common, vertical drain under certain conditions.

True False

22) Double sanitary tees can be used when the barrel of the fitting is at least one pipe size larger than the largest inlet.

True False

23) All of the following fixtures are considered to be suds-producing fixtures: bathtubs, kitchen sinks, dishwashers, laundries, and washing-machine standpipes.

True False

24) Fixture outlets connected to drainage piping within 8 feet of a vertical to horizontal change of direction of a stack containing suds-producing fixtures is prohibited, with two exceptions.

True False

25) The two exceptions mentioned in question 24 are as follows: single-family residences and stacks that receive the discharge from less than 3 stories of plumbing fixtures.

True False

This concludes the true-false test for zone 1, but there is more that you should know. The plumbing code rules in zone 1 are very different from some of the other codes, such as the code for zone 3. If you are a zone-1 plumber and didn't do well on some of the questions in the multiple-choice exam, it might not be your fault. It is next to impossible to give a mock exam that covers all the codes with equal effectiveness. To overcome this problem, consult your local code book. It may be faster and easier to refer to one of my other books, *National Plumbing Codes Handbook*, also published by McGraw-Hill.

In the *National Plumbing Codes Handbook*, I provide reminder notes at the back of each chapter to help readers determine the differences between the various codes. Some of the differences are substantial and should be investigated. This can be accomplished with the use of a local code book, but my *Handbook* makes the comparison work fast and easy to understand.

The plumbing code is a complicated piece of writing, but my *National Plumbing Codes Handbook* breaks down the cryptic language of the plumbing codes and puts it in terms any plumber can understand. Since I wrote the book, I am obviously biased, but I firmly believe the *Handbook* will make studying for your exam much simpler.

Whether you consult my *Handbook* or your own code book, be sure to check your local regulations to see what rulings apply in your area. My mock tests are written to help you improve your knowledge of the plumbing code, but it is your responsibility to understand your current, local regulations.

True-false test for zone 2

1) Underground piping within a building may not consist of aluminum pipe.

   True False

2) Pipe, which is subject to corrosion, that is installed underground must be protected when installed in corrosive soils.

   True False

3) Vertical vent stacks must be designed to control expansion and contraction.

True False

4) Joints on sewer piping must be root proof.

True False

5) When sanitary and storm drainage systems are installed on the same property, the two different drain systems can be installed side by side in a common trench.

True False

6) Old building sewers and building drains can always be used in conjunction with new plumbing.

True False

7) Horizontal building drainage piping with a diameter of 2 inches or less must be installed with a minimum grade of ⅛ inch per foot.

True False

8) Horizontal building drainage piping with a diameter of 3 inches or more must be installed with a minimum grade of ⅛ inch per foot.

True False

9) Sumps used to collect sewage must be tightly covered and vented.

True False

10) Both a check valve and gate valve are required in the installation of a sewer ejector.

True False

11) Sewer sumps that receive the discharge of more than four water closets must be equipped with two sewer pumps.

True False

12) The direct connection of a steam exhaust, blowoff, or drip pipe can be connected to the building drainage system in single-family homes only.

True False

13) Waste water discharged into the building drainage system must not exceed a temperature of 120 degrees Fahrenheit.

True False

14) A pumping system is said to contain a sump pump, pit, and discharge piping.

True False

15) Sump pumps may not be used to remove raw sewage.

True False

16) A sump pit for a sump pump must not be less than 18 inches in diameter and not less than 24 inches deep.

True False

17) Sump pits must be readily accessible.

True False

18) Sump pits must be fitted with removable covers.

True False

19) Pumps that connect to a drainage system must be connected with a sanitary tee fitting.

True False

20) Pumps that connect to a drainage system must be connected at least 10 feet from the base of any soil stack, waste stack, or fixture drain.

True False

21) Soil pipes installed outside of buildings must be protected from freezing temperatures.

True False

22) A sump for a sewage ejector may be constructed of cast iron.
True False

23) When conditions will not allow drains to be installed with the percentage of fall specified by the plumbing code, a lesser slope can be used so long as the computed velocity will not be less than 2 feet per second.
True False

24) Cast-iron sewers can be installed in filled or unstable ground.
True False

25) Borosilicate glass pipe can be used when installing piping for acid soil and waste drainage systems.
True False

True-false test for zone 3

1) Chemical waste drainage systems must be completely separated from sanitary drainage systems.
True False

2) A bathroom group consists of a lavatory, water closet, and bathtub.
True False

3) A check valve must be installed on the inlet pipe to a drainage sump.
True False

4) Grinder pumps or grinder ejectors that receive the discharge of water closets must be capable of handling 2½-inch solids.
True False

5) Sewage pumps and ejectors must discharge the contents of sewage sumps directly into a building sewer.
True False

6) All fittings used in drainage pipe must guide sewage and waste in the direction of flow.

True False

7) All drainage piping for future fixtures must terminate with an approved plug.

True False

8) Quarter bends may not be used for changes in direction that go from horizontal to vertical.

True False

9) Building subdrains must discharge into approved sumps that are gastight and vented.

True False

10) Minimum sump dimensions are 18 inches in diameter and 24 inches in depth.

True False

11) The capacity in gallons per minute (GPM) for a sewer pump that is fitted with a 2-inch discharge pipe is 21 GPM.

True False

12) The capacity in gallons per minute (GPM) for a sewer pump that is fitted with a 2½-inch discharge pipe is 21 GPM.

True False

13) The capacity in gallons per minute (GPM) for a sewer pump that is fitted with a 3-inch discharge pipe is 46 GPM.

True False

14) The discharge pipe for a sewage ejector must be fitted with a check valve between the pump and the point of discharge.

True False

15) A wye fitting can be used for changes in direction that go from horizontal to vertical, vertical to horizontal, and from horizontal to horizontal.

True False

16) A private water closet has a fixture-unit rating of 4.

True False

17) The minimum trap size for any fixture that is not referenced in the code book is 1½ inches.

True False

18) A drainage system that conveys sewage and excludes storm, surface, or ground water is called a sanitary system.

True False

19) Pipes with diameters of 3 inches or more can be installed with a grade of ⅛ inch per foot.

True False

20) A 4-inch trap has a fixture-unit rating of 6.

True False

21) Future fixtures that are roughed-in, but not set, are not counted when computing a fixture-unit load.

True False

22) Sanitary tee fittings can be used only when the change in direction is from vertical to horizontal.

True False

23) The storage period for collecting waste in a sewage sump shall not exceed 12 hours.

True False

24) Double sanitary tees are meant to receive the discharge from blowout fixtures.

True False

25) Fixtures with pumping action discharge may not discharge into double sanitary tees.

True False

You have now completed the tests on sanitary drainage systems. Turn to the answers at the end of the chapter and compare the correct answers with the answers you gave to the questions. If you answered many of the questions incorrectly, be sure to go back through your code book and find out why. Sanitary drainage systems are a big part of many plumbing exams, so you must learn this material thoroughly in order to do well on the real test.

Once you have checked your answers for these tests, move on to the next chapter and test your knowledge on the subject of indirect and special wastes.

Answers to multiple-choice exam

1) C
2) D
3) A
4) C
5) D
6) A
7) B
8) C
9) D
10) B
11) C
12) C
13) B
14) A
15) C
16) B
17) D
18) B
19) A
20) B
21) A
22) B
23) B
24) B
25) B
26) B
27) A
28) B
29) B
30) B
31) B
32) D
33) D
34) D
35) C
36) B
37) B
38) A
39) B
40) A
41) B
42) A
43) C
44) C
45) B
46) B
47) A
48) C
49) B
50) B
51) B
52) C
53) A
54) A
55) D
56) D
57) C
58) D
59) B
60) C
61) D
62) B
63) C
64) B
65) D
66) D
67) D
68) D
69) D
70) A
71) D
72) D
73) A
74) D
75) C
76) C
77) C
78) D
79) B
80) C
81) C
82) C
83) A
84) B
85) C

Answers to true-false test for zone 1

1) T	10) T	18) F
2) T	11) T	19) T
3) F	12) T	20) T
4) F	13) T	21) T
5) T	14) F	22) F
6) T	15) T	23) T
7) F	16) T	24) T
8) T	17) T	25) T
9) F		

Answers to true-false test for zone 2

1) F	6) F	11) F	16) T	21) T
2) T	7) F	12) F	17) F	22) T
3) T	8) T	13) F	18) T	23) T
4) T	9) T	14) T	19) F	24) T
5) T	10) T	15) T	20) T	25) T

Answers to true-false test for zone 3

1) T	10) T	18) T
2) F	11) T	19) T
3) F	12) F	20) T
4) F	13) T	21) F
5) F	14) T	22) F
6) T	15) T	23) T
7) F	16) T	24) F
8) F	17) T	25) T
9) T		

Table 7-1 **Fixture-Unit Requirements on Trap Sizes in Zone 1**

Pipe size	Fixture-unit discharge on stack from a branch	Total fixture-units allowed on stack
1½"	2	8
2"	6	24
3"	16*	60*
4"	90	500

*Note—no more than two toilets may be placed on a three-inch branch, and no more than six toilets may be connected to a three-inch stack.

Table 7-2 **Fittings Approved for Horizontal Changes in Zone 1**

1¼" trap	One fixture-unit
1½" trap	Three fixture-units
2" trap	Four fixture-units
3" trap	Six fixture-units
4" trap	Eight fixture-units

Table 7-3 **Fittings Approved for Vertical-to-Horizontal Changes in Zone 1**

Pipe size	Fixture-unit discharge on stack from a branch	Total fixture-units allowed on stack
1½"	3	4
2"	6	10
3"	20*	30*
4"	160	240

*Note—no more than two toilets may be placed on a three-inch branch, and no more than six toilets may be connected to a three-inch stack.

Fittings Approved for Horizontal-to-Vertical Changes in Zone 1 — Table 7-4

Forty-five-degree wye

Combination wye and eighth-bend

*Note—other fittings with similar sweeps may also be approved.

Minimum Drainage-Pipe Pitch in Zone 1 — Table 7-5

Forty-five-degree branches

Sixty-degree branches and offsets, if they are installed in a true vertical position

Stack-Sizing Tall Stacks in Zone 2 — Table 7-6

1¼" trap	One fixture-unit
1½" trap	Two fixture-units
2" trap	Three fixture-units
3" trap	Five fixture-units
4" trap	Six fixture-units

Stack-Sizing Table for Zone 2 — Table 7-7

Pipes under 3" in diameter	¼" to the foot
Pipes 3" or larger in diameter	⅛" to the foot

Horizontal-Branch Sizing Table in Zone 2 — Table 7-8

Pipes under 4" in diameter	¼" to the foot
Pipes 4" or larger in diameter	⅛" to the foot

Table 7-9

Minimum Drainage-Pipe Pitch in Zone 2

Pipe size	Pipe grade to the foot	Maximum number of fixture-units
2"	¼"	21
3"	¼"	42*
4"	¼"	216

*Note—no more than two toilets may be installed on a three-inch building drain.

Table 7-10

Fixture-Unit Requirements on Trap Sizes in Zone 3

Pipe size	Fixture-unit discharge on stack from a branch	Total fixture-units allowed on stack
1½"	2	4
2"	6	10
3"	20*	48*
4"	90	240

*Note—no more than two toilets may be placed on a three-inch branch, and no more than six toilets may be connected to a three-inch stack.

Table 7-11

Fixture-Unit Ratings in Zone 3

Pipe size	Fixture-unit discharge on stack from a branch	Total fixture-units allowed on stack
1½"	2	8
2"	6	24
3"	20*	72*
4"	90	500

*Note—no more than two toilets may be placed on a three-inch branch, and no more than six toilets may be connected to a three-inch stack.

Building-Drain Sizing Table for Zone 3

Table 7-12

Forty-five-degree wye

Sixty-degree wye

Combination wye and eighth-bend

Sanitary tee

Sanitary tapped-tee branches

*Note—cross fittings, like double sanitary tees, cannot be used when they are of a short-sweep pattern, however, double sanitary tees can be used if the barrel of the tee is at least two pipe-sizes larger than the largest inlet.

Stack-Sizing Table for Zone 3

Table 7-13

Pipes under 3" in diameter	1/4" to the foot
Pipes 3" to 6" in diameter	1/8" to the foot
Pipes 8" or larger in diameter	1/16" to the foot

Stack-Sizing Tall Stacks in Zone 3

Table 7-14

Bathtub	2
Shower	2
Residential toilet	4
Lavatory	1
Kitchen sink	2
Dishwasher	2
Clothes washer	3
Laundry tub	2

Table 7-15 **Minimum Drainage-Pipe Pitch in Zone 3***

Pipe size	Maximum number of fixture units on a horizontal branch
1¼"	1
1½"	3
2"	6
3"	20**
4"	160
6"	620

*Note—table does not represent branches of the building drain, and other restrictions apply under battery-venting conditions.

**Note—not more than two toilets may be connected to a single three-inch horizontal branch. Any branch connecting with a toilet must have a minimum diameter of three inches.

Table 7-16 **Fixture Unit Rating for Stacks**

Pipe size	Fixture-unit discharge on stack from a branch	Total fixture-units allowed on stack
1½"	2	8
2"	6	24
3"	16*	60*
4"	90	500

*Note—no more than two toilets may be placed on a three-inch branch, and no more than six toilets may be connected to a three-inch stack.

CHAPTER 8

Requirements for indirect waste piping and special wastes

NOT all plumbers concern themselves with learning all the code has to say about requirements for indirect waste piping and special wastes. These plumbers feel that indirect and special wastes don't affect them.

However, indirect and special wastes can play a significant role in the life of a plumber. For example, the standpipe for an automatic clothes washer is indirect waste. Another example of indirect waste is an airgap for a dishwasher. The point is, whether you are doing commercial or residential work, you are likely to work with some form of indirect or special waste.

The subject of indirect and special waste is a limited one. Unlike normal drainage work, there are relatively few rules and regulations that pertain to indirect and special wastes. However, the rules that do apply to these wastes are important.

Many indirect and special waste arrangements are required because of the discharge of potentially harmful substances. This fact alone makes the topic one with which every professional plumber should become acquainted.

Commercial plumbers do tend to work with indirect and special wastes more than residential plumbers, but all plumbers must learn the rules and regulations that pertain to these types of wastes. You can expect to see questions on the licensing exam about indirect and special wastes. If you want to pass the test, you'd better learn all you can about indirect and special wastes. With this in mind, let's move on to the multiple-choice exam on this subject.

Multiple-choice exam

1) The piping arrangement in which a drain from a fixture discharges indirectly into another fixture, such as when a clothes washer discharges into a laundry tub, at a point below the flood-level rim is known as:

A) An airgap
B) An airbreak
C) A receptor drain
D) Both A and C

2) Pipes that discharge into a drainage system, but which are not connected directly to the drainage system are called:

A) Direct wastes
B) Indirect wastes
C) Complex wastes
D) Special wastes

3) For a basin to be considered a swimming pool, it must have a depth of at least _____ feet at any point. Fill in the blank.

A) 2
B) 3
C) 4
D) 5

4) The unobstructed vertical distance through free atmosphere between the outlet of a waste pipe and the flood-level rim of a receptacle into which a waste pipe discharges is known as:

A) An airbreak
B) An airgap
C) A vertical gap
D) A free fall

5) Food-handling equipment must have its waste discharged through:

A) A direct connection
B) An indirect waste pipe
C) A grease interceptor
D) Both A and C

6) Equipment and fixtures used to store, prepare, or handle food must have their waste discharge through:

A) A direct connection
B) An indirect waste pipe
C) An interceptor
D) A check valve

7) An exception to the rule stated in question 6 is:

A) A dishwashing machine
B) A dishwashing sink
C) Both A and B
D) None of the above

8) When a device, such as a sterilizer, discharges potable water into a building drainage system, the discharge must be made through an indirect waste pipe by means of:

A) Gravity
B) An airgap
C) An airbreak
D) Both A and B

9) When a device, such as a process tank, discharges nonpotable water into a building drainage system, the discharge must be made through an indirect waste pipe by means of:

A) Gravity
B) An airgap
C) An airbreak
D) Both A and B

10) All indirect waste piping must discharge through:

A) An airgap or airbreak
B) A 2-inch or larger pipe
C) A filter
D) All of the above

11) Waste receptors and standpipes must be:

A) Trapped
B) Vented
C) Both A and B
D) Either A or B

12) All waste receptors and standpipes must connect to:

A) Storm-water drainage systems
B) Dry-well drainage systems
C) Building drainage systems
D) French drains

13) An indirect waste pipe that measures more than 2 feet in developed length, measured horizontally, must be:

A) Trapped
B) Vented
C) Trapped and vented
D) Marked as an indirect waste pipe

14) An indirect waste pipe that measures more than 4 feet in developed length must be:

A) Trapped
B) Vented
C) Trapped and vented
D) Marked as an indirect waste pipe

15) An airbreak is required between the indirect waste pipe and the _____ of the waste receptor or standpipe. Fill in the blank.

A) Airgap
B) Developed length
C) Trap seal
D) Flood-level rim

16) The airgap between an indirect waste pipe and the flood-level rim of the waste receptor must be a minimum of _____ time(s) the effective opening of the indirect waste pipe.

A) 1
B) 2
C) 4
D) None of the above

17) Equipment utilized to store food must discharge through an indirect waste pipe by means of:

A) An airgap
B) An airbreak
C) An air hose
D) An air relief

18) Which of the following is required to cover the waste outlet of waste receptors?

A) A removable strainer
B) A removable basket
C) Both A and B
D) Either A or B

19) Waste receptors are to be installed in:

A) Bathrooms
B) Kitchens
C) Ventilated areas
D) Restricted areas

20) Waste receptors must be:

A) Installed in bathrooms
B) Accessible
C) Readily accessible
D) At least 3 inches in diameter

21) Waste receptors may not be installed in:

A) Bathrooms
B) Toilet rooms
C) Both A and B
D) None of the above

22) Waste receptors must be installed to prevent:

A) Streaking
B) Splashing
C) Osmosis
D) Singular displacement

23) Which of the following types of waste may be discharged into the hub of a pipe?

A) Chemical wastes
B) Special wastes
C) Clear-water wastes
D) Any of the above

24) If the hub of a pipe is used as an indirect-waste receptor, the hub must extend at least _____ inches above a water-impervious floor. Fill in the blank.

A) 2
B) 3
C) 4
D) 6

25) When the hub of a pipe is used as an indirect-waste receptor, a _____ is not required. Fill in the blank with the most appropriate answer.

A) Discharge tube
B) Discharge relief vent
C) Strainer
D) Receptor

26) Standpipes require _____. Fill in the blank.

A) Strainers
B) Baskets
C) Individual traps
D) Screens

27) The minimum height for a standpipe is:

A) 12 inches
B) 16 inches
C) 18 inches
D) 48 inches

28) The maximum height for a standpipe is:

A) 31 inches
B) 32 inches
C) 42 inches
D) 48 inches

29) Standpipes with a diameter of 1½ inches that serve domestic, automatic clothes washers must not extend more than:

A) 18 inches
B) 28 inches
C) 30 inches
D) 32 inches

30) The traps of all standpipes must be:

A) Accessible
B) Readily accessible
C) Crown vented
D) Drum traps

31) Water being discharged into the drainage system of a building's plumbing must not exceed a temperature of:

A) 120 degrees Fahrenheit
B) 130 degrees Fahrenheit
C) 140 degrees Fahrenheit
D) 160 degrees Fahrenheit

32) Drains conveying water at a temperature in excess of that allowed for discharge into a drainage system directly must be piped into the drainage system by means of:

A) A cool-down chamber
B) An indirect-waste receptor
C) A baffled interceptor
D) None of the above

33) Which of the following are considered special wastes?

A) Corrosive liquids
B) Spent acids
C) Harmful chemicals
D) All of the above

34) Special wastes cannot be introduced into a plumbing system until the waste has been:

A) Diluted
B) Monitored
C) Screened
D) All of the above

35) Special wastes cannot be introduced into a plumbing system until the waste has been:

A) Neutralized
B) Manifested
C) Screened
D) All of the above

36) Chemical waste requires:

A) Its own drainage system
B) Its own vent system
C) Both A and B
D) None of the above

37) Commercial dishwashing machines must be connected to the drainage system in which of the following way(s):

A) Directly
B) Through a metallic piping material
C) Indirectly
D) Both A and B

38) The waste outlet for an indirect waste from food-handling equipment must terminate at least _____ inches above the flood-level rim of the receptor. Fill in the blank.

A) 2
B) 3
C) 4
D) 6

39) The drains, overflows, and relief vents from the water supply system or air-conditioning units must be connected through:

A) Direct waste connections
B) Indirect waste connections
C) Either A or B
D) Flexible connections

40) The maximum length of an indirect waste to vent shall not exceed:

A) 4 feet
B) 10 feet
C) 15 feet
D) 25 feet

41) Drinking fountains _____ be installed with indirect wastes. Fill in the blank.

A) Shall
B) Must
C) May
D) Will always

42) When the hub of a fitting is used as an indirect-waste receptor, the floor covering below the receptor must be:

A) Level
B) Pitched away from the drain
C) Impervious to water
D) Both B and C

43) Substances that might be harmful to the drains in a plumbing system are considered to be:

A) Indirect wastes
B) Direct wastes
C) Special wastes
D) Suspicious wastes

44) Chemical waste shall not be discharged into a drainage system until:

A) It has been treated in a manner approved by the plumbing code
B) It has been cooled to a temperature not exceeding 130 degrees Fahrenheit
C) It has been identified as a non-toxic chemical
D) All of the above

45) Special wastes might create:

A) Toxic fumes
B) Vertigo
C) Eczema
D) Both B and C

You have completed the multiple-choice exam that deals with indirect and special wastes. There is still a true-false test for you to take before you check your answers.

I want to remind you that there are differences in the plumbing codes for various parts of the country and that plumbing codes are amended from time to time. For these reasons the code in your area might generate answers that are different from those given for the multiple-choice questions. This should occur rarely, but it is possible. To be absolutely sure you have answered the questions correctly, refer to your local code book. My other book, *National Plumbing Codes Handbook*, might help you verify the code considerations in your region.

The questions found throughout these chapters have been derived from my personal knowledge and from reviewing code books that were current at the time this text and the *Handbook* were written. I cannot guarantee that there will not be alterations to the codes by the time you take the tests. I bring this point to light to make sure that you are prepared for the licensing exam. Now, let's move on to the true-false questions for your region.

True-false test for zone 1

1) Cooling and air-conditioning equipment may be separated from a drainage system with the use of an airbreak.

   True False

2) Food equipment must be separated from a drainage system with the use of an airgap.

   True False

3) When airgap drainage is used, the distance of the gap must not be less than 1 inch.

   True False

4) An airbreak may not terminate below the rim of a receptor.

   True False

5) Vents serving indirect-waste piping may be combined with the vents from a sewer once they have been extended 42 inches above the highest fixture's flood-level rim.

   True False

6) Indirect-waste pipes that are less than 15 feet in length are not required to be trapped.

   True False

7) Indirect-waste pipes that are less than 15 feet in length are not required to be vented.

   True False

8) All receptors for indirect waste must be readily accessible.

   True False

9) Automatic clothes washers used for domestic purposes may discharge into a trap that is below the floor where the appliance is situated.

   True False

10) A trap for a clothes washer must be roughed in at least 6 inches above the floor.

True False

11) A trap for a clothes washer must be roughed in at a height of no more than 18 inches above the floor.

True False

12) Indirect-waste piping may not be installed in a storeroom.

True False

13) The receptor for a clothes washer may be installed in a bathroom if the clothes washer is installed in the bathroom.

True False

14) Devices used as sterile equipment must be drained through an airbreak.

True False

15) Domestic dishwashers may be connected directly to a food waste disposer without the use of an airgap.

True False

16) Drinking fountains must be installed with indirect wastes.

True False

17) When piping is installed to handle chemical waste, the master plumber installing the piping must make and keep a permanent record of the location of the piping.

True False

18) The location of any vent used in conjunction with drainage for chemical waste must be recorded and kept permanently by the property owner.

True False

19) Rules and regulations for chemical wastes do not apply to small photographic darkrooms.

True False

20) Chemical waste may not be discharged into the ground.

True False

True-false test for zone 2

1) An indirect-waste pipe with a length of 2 feet or more must be trapped.

True False

2) An indirect-waste pipe with a length of 1 foot or less must be trapped.

True False

3) Appliances not regularly classed as plumbing fixtures, but which have drips or drainage outlets must be drained by indirect-waste pipes that discharge into open receptacles.

True False

4) The condensate or waste from air-conditioning equipment can be classified as a plumbing fixture under certain circumstances.

True False

5) An indirect-waste pipe may run up to 35 feet in length from its vent.

True False

6) Waste pipes for swimming pools are not allowed to be fitted with circulation pumps.

True False

7) The airgap between an indirect waste and a building drainage system must be at least twice the effective diameter of the drain being served.

True False

8) Waste receptors may not be installed in closets.

True False

9) Overflow pans for water heaters must discharge into the building drainage system through an indirect waste.

True False

10) The strainers and baskets used in conjunction with indirect-waste receptors must be constructed so as to prevent the passage of solids with diameters of ½ inch or larger.

True False

True-false test for zone 3

1) A waste receptor must be sized for the maximum discharge of all indirect-waste pipes served by it.

True False

2) Clear-water waste receptors are not required to have strainers installed.

True False

3) Potable clear-water waste may enter the drainage system through an airbreak.

True False

4) Chemical waste can be discharged into a sanitary drainage system once it has been treated in accordance with the plumbing code.

True False

5) Steam pipes may only be connected to a sanitary drainage system when the connection is made below ground.

True False

6) Waste water entering a sanitary drainage system must have a temperature of no more than 120 degrees Fahrenheit.

True False

7) The traps of all standpipes must be accessible for rodding.

True False

8) The drains of all standpipes must be accessible for rodding.

True False

9) An airbreak must be provided between an indirect-waste pipe and the trap seal of the waste receptor or standpipe.

True False

10) The airgap between an indirect-waste pipe and the flood-level rim of the waste receptor must be at least twice the effective opening of the indirect waste pipe.

True False

You have now completed all the tests on indirect and special wastes. The rules that pertain to these wastes are necessary to protect public health and safety. Don't allow yourself to take lightly the use of indirect and special wastes. Go to the end of the chapter and check your answers against the correct answers. Whenever you answered incorrectly, consult your code book to confirm the correct answer and to learn why the answer is what it is.

When you have finished grading and reviewing your answers to these tests, move on to the next chapter that deals with the use of cleanouts in drainage systems.

Answers to multiple-choice exam

1) B	8) B	15) C	22) B
2) B	9) C	16) B	23) C
3) A	10) A	17) A	24) A
4) B	11) C	18) D	25) C
5) B	12) C	19) C	26) C
6) B	13) A	20) C	27) C
7) C	14) A	21) C	28) C

29) C	34) A	38) A	42) C
30) A	35) A	39) B	43) C
31) C	36) C	40) D	44) A
32) B	37) C	41) C	45) A
33) D			

Answers to true-false test for zone 1

1) T	8) T	15) F
2) T	9) F	16) F
3) T	10) T	17) F
4) F	11) T	18) T
5) F	12) T	19) T
6) F	13) T	20) T
7) T	14) F	

Answers to true-false test for zone 2

1) T	5) F	8) T
2) F	6) F	9) T
3) T	7) T	10) T
4) T		

Answers to true-false test for zone 3

1) T	5) F	8) T
2) T	6) F	9) T
3) F	7) T	10) T
4) T		

CHAPTER 9

Drainage cleanout requirements

THE rules for drainage cleanouts are not extensive. They are, however, important to any plumber who has ever had to clean a drain. They are also important when you take a licensing exam or attempt to pass a plumbing inspection in the field. Cleanouts are typically taken for granted, until they are needed. You must not overlook cleanout requirements when you prepare for the exam or install a drainage system.

I see code violations on a weekly basis. Most any service plumber encounters a number of situations where the plumbing code was not followed. Of all the violations I see, cleanouts tend to be one of the most prominent.

If you have ever had to snake a drain where there was no cleanout or where the cleanout was installed without adequate clearance, you can appreciate the intent behind the plumbing code. For those readers who have not yet dealt with such situations, let me just say that a day will come when you will wish the plumber before you had complied with the code when he or she installed the drainage system with which you have to work.

This is not a long chapter, but it does cover the rules and regulations that pertain to cleanouts in great depth. Let's begin this coverage with a multiple-choice test.

Multiple-choice exam

1) Which of the following constitute a cleanout?

A) A removable plug or cap
B) A removable fixture trap
C) A water closet
D) All of the above

2) An access opening in a drainage system that is used for the removal of obstructions is:

A) An interceptor
B) A cleanout
C) An airbreak
D) A continuous waste

3) When installing a horizontal drain that has a diameter of 4 inches or less, you must install cleanouts at intervals of not more than:

A) 25 feet
B) 35 feet
C) 50 feet
D) 75 feet

4) When installing a horizontal drain that has a diameter larger than 4 inches, you must install cleanouts at intervals of not more than:

A) 50 feet
B) 75 feet
C) 100 feet
D) 125 feet

5) All building sewers must have cleanouts that are no more than _____ feet apart. Fill in the blank.

A) 50
B) 75
C) 100
D) 125

6) Cleanouts must be installed at each change of direction in a building drain or horizontal drain when the change of direction is _____ degrees or more. Fill in the blank.

A) 16
B) 22½
C) 36
D) 45

7) When more than one change in direction occurs in a run of piping, only one cleanout is required for each ______ feet of developed length of drainage piping. Fill in the blank.

A) 20
B) 40
C) 50
D) 75

8) All cleanouts must be provided with:

A) Brass plugs
B) Brass caps
C) Countersunk heads
D) Access

9) A cleanout is required at the base of each:

A) Fixture drain
B) Waste stack
C) Trap
D) Both A and C

10) A cleanout is required at the base of each:

A) Kitchen sink
B) Water closet
C) Soil stack
D) Both A and C

11) A cleanout is required at the junction between a building drain and a:

A) Horizontal branch
B) Trap arm
C) Building sewer
D) All of the above

12) Cleanouts for underground drains and sewers outside of buildings must be extended to:

A) Ground level
B) A point 4 inches above the ground level
C) A point 4 inches below the ground level
D) A point 2 inches above the ground level

13) Cleanouts for underground drains installed inside a building must be extended to:

A) Floor level
B) A point 4 inches above the floor level
C) A point 4 inches below the floor level
D) A point 2 inches above the floor level

14) Cleanouts installed for pipes with diameters of 4 inches or less must be:

A) One half the size of the drainage pipe being served
B) Twice the size of the drainage pipe being served
C) The same size as the pipe being served
D) Made of plastic

15) Cleanouts installed at the junction of a building drain and building sewer may be installed:

A) Inside the building
B) Outside the building
C) Either A or B
D) Neither A nor B

16) A cleanout is not required at the junction of a building drain and building sewer if there is a cleanout installed in a 3-inch, or larger, stack within _____ feet of the junction.

A) 5
B) 8
C) 10
D) 15

17) A drainage pipe with a diameter of 5 inches is required to have cleanouts of what size?

A) 3 inch
B) 4 inch
C) 5 inch
D) 6 inch

18) A drainage pipe with a diameter of 6 inches is required to have cleanouts of what size?

A) 3 inch
B) 4 inch
C) 5 inch
D) 6 inch

19) A drainage pipe with a diameter of 8 inches is required to have cleanouts of what size?

A) 3 inch
B) 4 inch
C) 5 inch
D) 6 inch

20) A drainage pipe with a diameter of 10 inches, or more, is required to have cleanouts of what size?

A) 4 inch
B) 6 inch
C) 8 inch
D) 10 inch

21) When a building sewer has a diameter of 10 inches or more, which of the following is required?

A) 8-inch cleanouts
B) Manholes
C) Thermal saddles
D) A map of the piping system

22) Underground building drains with diameters of 10 inches or more must be equipped with:

A) 8-inch cleanouts
B) Manholes
C) Either A or B
D) Both A and B

23) Manholes must be equipped with:

A) Vents
B) Overflow drains
C) Gas-tight covers
D) All of the above

24) When manholes are installed along building sewers that have diameters of 10 inches or more, the manholes must be no more than _____ feet apart. Fill in the blank.

A) 100
B) 175
C) 200
D) 400

25) When installing manholes for building sewers with diameters of 10 inches or more, one manhole must be installed for each _____. Fill in the blank.

A) Change in direction
B) Change in elevation
C) Either A or B
D) Both A and B

26) When installing manholes for building sewers with diameters of 10 inches or more, one manhole must be installed for each _____. Fill in the blank.

A) Change in grade
B) Change in size of the building sewer
C) Either A or B
D) Both A and B

27) When manholes are used in flood-hazard zones or high-hazard zones, they must be installed so that _____. Fill in the blank.

A) They are elevated to or above the base flood elevation
B) They are sealed to resist hydrostatic and hydrodynamic loads and stresses, including the effects of buoyancy during the occurrence of flooding to the base flood elevation
C) Either A or B
D) Both A and B

28) Cleanouts should either open in the direction of the flow of the drainage pipe or _____. Fill in the blank.

A) At 45-degree angles thereto
B) At right angles thereto
C) At 180 degrees thereto
D) Any of the above

29) A clearance of 18 inches is required for all cleanouts installed on pipes with diameters of:

A) 4 inches or more
B) 6 inches or more
C) 6 inches or less
D) 8 inches or less

30) A clearance of 36 inches is required for all cleanouts installed on pipes with diameters of:

A) 6 inches or more
B) 8 inches or more
C) 10 inches or less
D) All of the above

31) Cleanout plugs must not be:

A) Made of plastic
B) Concealed
C) Covered permanently
D) Painted

32) Upon approval, cleanout openings may be used for:

A) The installation of new fixtures
B) The installation of floor drains
C) Both A and B
D) None of the above

33) Anytime a cleanout is concealed, provisions must be made so that the cleanout is:

A) Readily accessible
B) Accessible
C) Labeled
D) Discontinued

34) For new work, cleanout plugs must be made of:

A) Heavy brass
B) Plastic
C) Either A or B
D) None of the above

35) For new work, cleanout plugs must have a minimum thickness of:

A) ⅛ inch
B) ¼ inch
C) ½ inch
D) ¾ inch

36) For new work, cleanout plugs must have:

A) A raised nut
B) A recessed socket
C) Either A or B
D) Neither A nor B

37) When it becomes necessary to repair a cleanout, which of the following can be used as a cleanout plug?

A) A heavy brass plug
B) A heavy lead plug
C) Either A or B
D) None of the above

38) Cleanouts installed in pipes with diameters of 3 inches or less must be installed at intervals not to exceed:

A) 25 feet
B) 50 feet
C) 75 feet
D) 100 feet

39) Cleanouts installed in pipes with diameters of 4 inches or more must be installed at intervals not to exceed:

A) 25 feet
B) 50 feet
C) 75 feet
D) 100 feet

40) Line cleanouts that may be rodded both ways should be installed:

A) In pipes with diameters of 4 inches or more
B) In pipes with diameters of 3 inches or less
C) Whenever practical
D) Only with written permission

41) Cleanouts smaller than 3 inches must be installed so that there is at least _____ inches of clearance for rodding. Fill in the blank.

A) 6
B) 12
C) 18
D) 24

42) For building sewers with diameters of 8 inches or more, manholes must be provided at intervals not to exceed _____ feet. Fill in the blank.

A) 100
B) 200
C) 300
D) 400

43) Cleanouts on the seal of a trap must be made watertight and gastight with:

A) Minimum pressure
B) A threaded cleanout plug and approved washer
C) A nonacid pipe sealant
D) No more than 4 turns

44) Countersunk heads may be used on cleanouts when raised heads might present a:

A) Installation difficulty
B) Problem
C) Hazard
D) More costly installation

45) When a hubless blind plug is used as a cleanout, the complete coupling and plug must be:

A) Accessible
B) Readily accessible
C) Visible
D) Labeled

This completes the multiple-choice exam on cleanouts. You will probably notice a little repetition in some of the questions. When you check the answers at the back of the chapter, you will see that the repetitious questions don't share the same answer. This is because different codes have different rulings. In such cases, you should get only one of the two similar questions right. Don't count the missed answer against yourself when you score the test. To confirm the correct answer for your region, consult your local code book or my *National Plumbing Codes Handbook*.

As we move to the true-false questions for the three primary regions, you will see more differences between the codes. However, in the true-false exams, the questions are keyed to specific regions so there will not be questions with conflicting answers. Let's start with the test for zone 1.

True-false test for zone 1

1) Excluding exceptions, cleanouts must be installed in horizontal drainage pipe at intervals of not more than 100 feet.

True   False

2) Cleanouts may be omitted from horizontal drain lines that do not exceed 10 feet in length.

True   False

3) Cleanouts are not required on any horizontal drainage piping that is above the first floor of a building.

True   False

4) When an approved two-way cleanout fitting is installed inside a building near the connection between the building drain and building sewer, there is no need to install an upper-end terminal cleanout.

True False

5) When an approved two-way cleanout fitting is installed outside a building at the lower end of the building drain and is extended to grade, there is no need to install an upper-end terminal cleanout.

True False

6) With the exception of a building drain and its horizontal branches, cleanouts are not required on any pipe above the first floor of a building.

True False

7) Cleanouts are not required on horizontal drains that are less than 5 feet in length, unless such drains serve sinks or urinals.

True False

8) Cleanouts are not required on any horizontal drain that is installed on a slope of 72 degrees or less from the vertical angle.

True False

9) When a horizontal drain changes direction with an angle exceeding 135 degrees, an additional cleanout must be installed.

True False

10) Cleanouts for interceptors must be installed within the interceptor.

True False

11) All cleanouts must be installed above floor level or above ground level.

True False

12) Cleanouts can be concealed by removable cover plates.

True False

13) Cleanouts with diameters of 2 inches or less must be installed to provide a minimum clearance of 12 inches for rodding.

True False

14) Cleanouts for pipes with diameters larger than 2 inches must be provided with a minimum clearance of 24 inches for rodding.

True False

15) Under-floor cleanouts in residential dwellings must be located no more than 15 feet from any access door, trap door, or crawl hole.

True False

16) Cleanouts, excluding exceptions, should be installed vertically above the flowline of the pipe they serve.

True False

17) Pipes with diameters of 3 inches must be fitted with cleanouts that allow 18 inches of clearance for rodding.

True False

18) Cleanouts are not required on drains with diameters of 1½ inches or less.

True False

19) Unless a cleanout is installed under an approved cover plate, the cleanout must be readily accessible.

True False

20) Cleanouts are required whenever a horizontal pipe has an aggregate change in direction of 45 degrees or more.

True False

True-false test for zone 2

1) Cleanout caps must be watertight and gastight.

True False

2) Nylon plastics are acceptable as an alternative material.

True False

3) The bodies of cleanout ferrules must conform in thickness to that required for pipe and fittings of the same material.

True False

4) Lead plugs are prohibited from use in the repair of cleanouts.

True False

5) Excluding exceptions, every horizontal drain pipe must be fitted with a cleanout at the upstream end of the pipe.

True False

6) When P-traps are connected to the drainage piping with slip joints or ground-joints connections, they eliminate the need for an upstream cleanout for that drainage branch.

True False

7) When a floor drain, shower drain, or tub drain discharges into a P-trap, and when the fixture is fitted with a removable strainer, an upstream cleanout is not required.

True False

8) When a residential washing machine discharges through an S-trap, an upstream cleanout is not required.

True False

9) Water closets serve as a replacement for upstream cleanouts on the closet fixture branch only.

True False

10) When a residential washing machine discharges through a P-trap, an upstream cleanout is not required.

True False

11) Cleanouts are required near the junction of a building drain and building sewer.

True False

12) When a test tee or cleanout is installed in a vertical pipe above the flood-level rim of fixtures that a horizontal pipe serves and the test tee or cleanout is not more than 4 feet above the floor level, an upstream cleanout is not required.

True False

13) Cleanouts are required at maximum intervals of 50 feet for drainage pipes with diameters of 3 inches or less.

True False

14) Cleanouts for 4-inch pipe must be spaced at intervals not exceeding 75 feet.

True False

15) A 1¼-inch P-trap may serve as a cleanout for a 1½-inch pipe.

True False

16) A 1½-inch P-trap may serve as a cleanout for a 2-inch pipe.

True False

17) A 2-inch P-trap may serve as a cleanout for a 3-inch pipe.

True False

18) An 8-inch building drain may be served by a 4-inch cleanout.

True False

19) When a two-way cleanout is installed near the junction of a building drain and building sewer, it shall not be necessary to install an upstream cleanout.

True False

20) Cleanouts shall not be used for the installation of new fixtures or floor drains without the expressed, written permission of a code official.

True False

True-false test for zone 3

1) Cleanouts for building sewers must not be located more than 150 feet apart.

   True False

2) A removable plug or cap is considered a cleanout.

   True False

3) Cleanouts for pipes with diameters of 4 inches or less must be of the same size as the pipe being served.

   True False

4) A 4-inch cleanout can be used with a 5-inch pipe or a 6-inch pipe.

   True False

5) Cleanouts for underground piping must extend to the finished grade level or the finished floor level.

   True False

6) Brass cleanout plugs are prohibited.

   True False

7) Manholes, when required, must be installed at intervals not exceeding 400 feet.

   True False

8) Cleanouts must be installed at the base of each waste or soil stack.

   True False

9) Cleanouts at the junction of building drains and building sewers may be installed inside a building or outside a building.

   True False

10) Drainage pipes with diameters of 10 inches or more can be served by 6-inch cleanouts.

   True False

11) If a cleanout is concealed, there must be sufficient access to allow for the removal of the cleanout and the rodding of the drain.

True False

12) Building drains with diameters of 10 inches or more must be equipped with manholes.

True False

13) The rodding clearance required for pipes with diameters of 6 inches or less is 18 inches.

True False

14) Cleanout plugs that are to be concealed in finished walls may be concealed with plaster or some other breakable substance.

True False

15) If a cleanout is used for the addition of new fixtures, another cleanout must be installed of equal access and capacity.

True False

16) Drains with diameters of 8 inches or more must be equipped with cleanouts that have a minimum of 32 inches of clearance for rodding.

True False

17) 18 inches is equivalent to 457 millimeters.

True False

18) Cleanouts must be readily accessible.

True False

19) 400 feet is equal to 122 meters.

True False

20) A removable fixture trap may serve as a cleanout.

True False

Well, we have come to the end of yet another series of tests. Your knowledge of cleanouts has been examined, and it is up to you to check your answers with the correct answers at the end of the chapter. Compare your work with the correct answers and research any mistakes you made. When you have completed this task, move on to chapter 10, and we will explore traps, interceptors, and backwater valves.

Answers to multiple-choice exam

1) D	13) A	24) D	35) A
2) B	14) C	25) D	36) C
3) C	15) C	26) D	37) C
4) C	16) C	27) C	38) B
5) C	17) B	28) B	39) C
6) D	18) B	29) C	40) C
7) B	19) D	30) B	41) B
8) D	20) C	31) C	42) C
9) B	21) B	32) C	43) B
10) C	22) C	33) B	44) C
11) C	23) C	34) C	45) A
12) A			

Answers to true-false test for zone 1

1) T	8) T	15) F
2) F	9) T	16) T
3) F	10) F	17) T
4) T	11) F	18) F
5) T	12) T	19) T
6) T	13) T	20) F
7) T	14) F	

Answers to true-false test for zone 2

1) T	8) F	15) T
2) T	9) T	16) T
3) T	10) T	17) F
4) F	11) T	18) T
5) T	12) T	19) T
6) T	13) T	20) T
7) T	14) F	

Answers to true-false test for zone 3

1) F	8) T	15) T
2) T	9) T	16) F
3) T	10) F	17) T
4) T	11) T	18) F
5) T	12) F	19) T
6) F	13) T	20) T
7) T	14) T	

CHAPTER 10

Traps, interceptors, and backwater valves

THIS chapter is going to test your knowledge of traps, interceptors, and backwater valves. There are many circumstances that can affect the trap that is used with a particular fixture. A lot of plumbers become confused when asked to explain the difference between a trap and an interceptor. While traps are not complicated devices, installing them according to the plumbing code gives some plumbers trouble.

For example, there sometimes is doubt as to whether an S-trap can be replaced with another S-trap or whether a different type of trap is required. Drum traps are common in Maine and rare in Virginia. Why is this? The answer is the venting systems required in the two different states.

A lot of plumbers don't even know what backwater valves are. It's true that backwater valves are not used on a regular basis by average plumbers, but that doesn't mean a plumber isn't required to understand them for the licensing exam.

Do you know when a backwater valve should be installed? Do you know what a backwater valve is? Does a backwater valve serve a water distribution system or a drainage system? If you can't answer all of these questions correctly at this time, don't worry, this chapter will enable you to learn about and understand backwater valves.

Most any plumber with a little experience knows what a P-trap is, but a much smaller percentage of plumbers can describe a crown-vented trap. Do you know? How about a bell trap. Can you describe what it is or where it is used? Are S-traps legal in new installations?

As you can see, many questions about traps can come up, and we've barely scratched the surface. Since all plumbing fixtures are usually trapped, it is important to know what traps to use, when to use them, and where to install them. Refer to Tables 10-1 through 10-4 at the end of this chapter for trap sizes by zone.

Like the last chapter, this chapter is not very long. The subject of traps, interceptors, and backwater valves doesn't take up a lot of space in a code book, and it won't consume a large number of pages in this book. This doesn't mean, however, that you can look past this chapter and move on to more interesting topics.

Traps, interceptors, and backwater valves play important roles in plumbing systems, and they deserve your attention and respect. There could be enough questions on these plumbing system elements to have an adverse effect on your overall score.

One thing too many potential plumbers fail to understand is the way that an exam can be worded to make a simple subject complicated. Some plumbing exams can have the best plumbers scratching their heads. I'm warning you, don't take any part of the plumbing code for granted.

If you make too many mistakes on your licensing exam, you will not get the license that you need to enjoy better working conditions and more money. Treat every question on your exam as if it stood between you and your plumber's license. Now, let's move on to the multiple-choice exam for this chapter.

Multiple-choice exam

1) A device, fitting, or assembly of fittings installed in a building drain for the purpose of preventing air circulation between the drainage system of the building and the building sewer is known as:

   A) A backwater valve   C) A separator
   B) An interceptor   D) A building trap

2) A device, fitting, or assembly of fittings installed to provide a liquid seal that prevents the emission of sewer gas, without materially affecting the flow of sewage or waste water through the device is known as:

A) An air chamber
B) A trap
C) A gas interceptor
D) An approved obstruction

3) The vertical distance between the crown weir and the top of the dip of a trap is known as:

A) An airgap
B) An airbreak
C) The critical level
D) The trap seal

4) A device that is designed and installed to separate and retain for removal deleterious, hazardous, or undesirable matter from normal wastes, while permitting normal sewage or wastes to discharge into the drainage system by gravity is known as:

A) A backwater valve
B) Interceptor
C) Drum trap
D) Screener

5) Excluding exceptions, every plumbing fixture must be equipped with:

A) Cut-off valves
B) A water-sealed trap
C) A gas-sealed trap
D) A P-trap

6) When a trap is installed to serve a plumbing fixture, the trap must be installed as close as possible to:

A) The floor
B) The trap arm
C) The vent
D) The fixture outlet

7) The vertical distance from the fixture outlet to the trap weir must not exceed:

A) 6 inches
B) 12 inches
C) 24 inches
D) 30 inches

8) No plumbing fixture is allowed to be:

A) Served by a continuous waste
B) Trapped with a drum trap
C) More than 20 inches from its vent
D) Double trapped

9) The trap of a kitchen sink may not accept the waste of:
A) A dishwasher
B) A garbage disposer
C) A continuous waste
D) Clothes washer

10) The drainage discharge from a laundry tub must not discharge into:
A) A P-trap
B) A sanitary drainage system
C) A septic tank
D) Kitchen-sink trap

11) Combination plumbing fixtures, such as double-bowl kitchen sinks, are allowed to share a single trap when the compartments of the sink bowls are no more than _____ inches deeper than the others. Fill in the blank.
A) 4
B) 6
C) 8
D) 12

12) Combination plumbing fixtures, such as double-bowl kitchen sinks, are allowed to share a single trap when the waste outlets are not more than _____ inches apart. Fill in the blank.
A) 12
B) 15
C) 30
D) 36

13) A single trap may be used to accommodate up to _____ single-compartment sinks that are adjacent to each other and in the same room. Fill in the blank.
A) 2
B) 3
C) 4
D) 5

14) When a single trap is used to accept the waste from multiple single-compartment sinks, as described in the last question, the trap must be:
A) Oversized
B) Centrally located
C) Metallic
D) Readily accessible

15) Fixture traps must not have:
A) Plastic bodies
B) Slip-joint connections
C) Interior partitions
D) Rough-brass finishes

16) As a plumber, you have been assigned the task of connecting a 3-bay commercial sink to the drainage system. The fixture outlets have diameters of 1½ inches. The trap arm and drain for the fixture has a diameter of 2 inches. Which of the following traps is prohibited under these circumstances?

A) A P-trap
B) A 1½-trap
C) A 2-inch trap
D) A 3-inch trap

17) Drum traps are prohibited, except:

A) For use with lavatories
B) For use with floor drains
C) For use with condensate drains
D) Where approved

18) Traps must be set level with respect to the:

A) Position of the fixture being served
B) Position of the floor beneath the trap
C) Trap seal
D) Direction of flow

19) Which of the following types of traps are prohibited?

A) S-traps
B) Crown-vented traps
C) All of the above
D) None of the above

20) Which of the following types of traps are prohibited?

A) P-traps
B) Crown-vented traps
C) All of the above
D) None of the above

21) Which of the following types of traps are prohibited?

A) Bell traps
B) Traps that depend on moving parts to maintain a seal
C) All of the above
D) None of the above

22) Which of the following types of traps are prohibited?

A) Separate fixture traps that depend on interior partitions for the seal
B) Traps that depend on moving parts to maintain a seal

C) All of the above
D) None of the above

23) Each fixture trap is required to have a liquid seal with a minimum depth of:

A) 2 inches
B) 3 inches
C) 4 inches
D) 6 inches

24) Each fixture trap is required to have a liquid seal with a maximum depth of:

A) 2 inches
B) 3 inches
C) 4 inches
D) 6 inches

25) Traps installed in unheated areas must be protected against:

A) Condensation
B) Freezing
C) Vandalism
D) All of the above

26) Under normal conditions, which of the following cannot be installed without permission from a code official?

A) Drum traps
B) Building traps
C) Both A and B
D) None of the above

27) When a building trap is installed, it must be provided with:

A) A cleanout
B) A relief vent or fresh-air intake
C) Both A and B
D) Either A or B

28) When an acid-resisting trap is installed underground, the trap must be embedded in concrete that extends _____ inches beyond the bottom and sides of the trap. Fill in the blank.

A) 2
B) 4
C) 6
D) 12

29) When an opening is left in a concrete slab for the connection of a fixture trap, such as a bathtub installed on a concrete slab, the opening must be:

A) Corrosion resistant
B) Insect proof
C) Vermin proof
D) All of the above

30) When a building trap is installed and a relief vent is provided for it, the vent must have a diameter that is at least _____. Fill in the blank.

A) One-half the size of the drain it is connected to
B) Equal to the size of the drain it is connected to
C) 3 inches in size
D) None of the above

31) Which of the following substances warrant the use of an interceptor?

A) Oil
B) Grease
C) Sand
D) All of the above

32) Grease interceptors are not required for:

A) Motel kitchens
B) Individual dwelling units
C) Correctional facilities
D) Restaurants

33) Grease interceptors are not required for:

A) Private living quarters
B) Individual dwelling units
C) Both A and B
D) Correctional facilities

34) All interceptors must be rated for:

A) Volume
B) Efficiency
C) Place of manufacture
D) All of the above

35) A fresh-air intake or vent for a building trap must be:

A) Carried above grade
B) 3 inches in diameter
C) Made of plastic pipe
D) None of the above

36) Which of the following locations would require the installation of a separator?

A) An apartment building
B) A bank building
C) A single-family residence
D) A repair garage with grease racks

37) Which of the following cannot be connected to a grease interceptor?
A) Hotel kitchens
B) Factory cafeterias
C) Restaurants
D) Food waste grinders

38) Grease interceptors must be equipped with devices to control the rate of:
A) Air flow
B) Water flow
C) Heat transference
D) All of the above

39) Which of the following types of waste are not allowed to be discharged into or through an interceptor or separator?
A) Flammable liquids
B) Oil
C) Wastes that do not require treatment
D) Grease

40) Oil separators must have a depth of at least _____ feet below the invert of the discharge drain. Fill in the blank.
A) 2
B) 4
C) 5
D) 6

41) The outlet opening of an oil separator must have a water seal of not less than _____ inches. Fill in the blank.
A) 2
B) 6
C) 12
D) 18

42) Separators that are installed in areas where automobiles are serviced must have a minimum capacity of _____ cubic feet for the first 100 square feet of area to be drained. Fill in the blank.
A) 4
B) 6
C) 8
D) 12

43) Separators, as described in question 42, must be enlarged at a rate of _____ cubic inches for each additional 100 square feet of area to be drained. Fill in the blank.
A) 4
B) 6
C) 8
D) 12

44) Sand interceptors must be installed so that they are:

A) Accessible
B) Readily accessible
C) Concealed
D) None of the above

45) Sand interceptors must have water seals with depths of not less than:

A) 6 inches
B) 10 inches
C) 14 inches
D) 18 inches

46) Filters used in commercial laundries must be sized to prohibit the entrance of articles with diameters of _____ or larger from entering the drainage system. Fill in the blank.

A) ¼ inch
B) ½ inch
C) ¾ inch
D) 1 inch

47) Commercial laundries must be equipped with an interceptor with a _____ or similar device that is removable for cleaning. Fill in the blank.

A) Water seal
B) Gas seal
C) Wire basket
D) Brush

48) The interceptors used in bottling establishments must be capable of separating _____ from other solids before discharging the waste into the drainage system. Fill in the blank.

A) Bottle caps
B) Broken glass
C) Paper labels
D) Bottled fluids

49) Separators used in slaughterhouses must be capable of preventing _____ from entering the drainage system. Fill in the blank.

A) Feathers
B) Entrails
C) Both A and B
D) None of the above

50) Backwater valves must be installed so that they are:

A) Accessible
B) Readily accessible
C) Not concealed
D) All of the above

51) If an interceptor or separator is subject to a loss of a trap seal, the device must be:

A) Drained of all air
B) Vented
C) Inspected regularly
D) All of the above

52) Interceptors must be installed so that they are:

A) Accessible
B) Readily accessible
C) Concealed
D) All of the above

53) Separators must be installed so that they are:

A) Accessible
B) Readily accessible
C) Concealed
D) All of the above

54) Interceptors must be cleaned:

A) Periodically
B) Every Monday
C) At the end of each work week
D) Once a month

55) Separators must be cleaned:

A) Periodically
B) Every week
C) At the end of each work week
D) Once a month

56) A reversal in flow of a drainage system is known as:

A) Back pressure
B) Back flow
C) Siphonic action
D) Any of the above

57) Backwater valves must be installed when plumbing fixtures are subject to back flow from:

A) An overflowing fixture
B) A building drain
C) A building sewer
D) A public sewer

58) All bearing parts of backwater valves must be:

A) Constructed of brass
B) Corrosion resistant
C) Both A and B
D) Either A or B

59) The diameter of a backwater valve when in its full-open position must be:

A) No less than ½ the diameter of the building drain
B) No less than the diameter of the pipe in which it is installed
C) No less than 1 pipe size larger than the pipe it serves
D) At least 2 inches

60) Baskets used in interceptors for commercial laundries must be:

A) Made of wire
B) Made of nylon mesh
C) Removable
D) Recyclable

You have now completed the multiple-choice test on traps, interceptors, separators, and backwater valves. Were the questions a bit more complex than what you expected? I suspect they were. As you can see, there is quite a bit to learn about these devices. If you underestimated the importance of traps, interceptors, separators, and backwater valves, be glad you did so on this mock examine rather than on the licensing exam. Now, proceed directly to the true-false test for your region.

True-false test for zone 1

1) All plumbing fixtures, except those with integral traps, must be trapped with an approved, waterseal trap.

True False

2) Not more than one trap is permitted on a trap arm.

True False

3) No more than two single-compartment sinks may discharge into a common trap with the use of a continuous waste.

True False

4) Commercial garbage disposers may not be installed with any sink served by a single trap.

True False

5) All commercial garbage disposers must be trapped individually.

True False

6) An automatic clothes washer can only discharge into the trap of a kitchen sink when the clothes washer is installed adjacent to the kitchen sink.

True False

7) The vertical distance between a fixture outlet and the trap weir cannot exceed 30 inches.

True False

8) Every plumbing trap, excluding exceptions, must be vented.

True False

9) A cleanout is required in a trap arm if the pipe has a diameter of less than 3 inches when the pipe takes a turn of 90 degrees or more.

True False

10) A 1¼-inch trap arm cannot extend more than 30 inches from its vent.

True False

11) A 1½-inch trap arm cannot extend more than 42 inches from its vent.

True False

12) A 2-inch trap arm cannot extend more than 42 inches from its vent.

True False

13) A 3-inch trap arm cannot extend more than 70 inches from its vent.

True False

14) A 4-inch trap arm cannot extend more than 10 feet from its vent.

True False

15) All traps, except those used for interceptors and similar devices, must be self-cleaning.

True False

16) Traps constructed of brass tubing may be used to receive domestic sewage, but the gauge of the metal shall not be less than 20 B&S Gauge.

True False

17) The trap for a fixture drain shall not be smaller than the trap arm it is connected to.

True False

18) S-traps are prohibited.

True False

19) Bell traps are allowed only for use with floor drains.

True False

20) Crown-vented traps are prohibited.

True False

21) Floor drains that are connected directly to a drainage system and that see seldom use must be equipped with a trap primer.

True False

22) Trap primers must be readily accessible.

True False

23) The plumbing code recognizes interceptors as clarifiers.

True False

24) All interceptors must be readily accessible.

True False

25) Interceptors that are installed in such a way as to require the use of a ladder for access are in violation of the plumbing code.

True False

26) All private and public auto wash racks or floors used for cleaning machinery or machine parts must be protected against storm or surface water.

True False

27) Private auto wash racks and floors used for the cleaning of machinery or machine parts are not required to discharge their drainage into an interceptor.

True False

28) Grease traps may not have flow-rates in excess of 50 gallons per minute.

True False

29) The flow-rate for a grease trap may not be less than 20 gallons per minute.

True False

30) Flow-control devices used with grease traps may not have adjustable or removable parts.

True False

True-false test for zone 2

1) Interceptors and separators must be installed when, in the opinion of a code official, they are necessary.

True False

2) The size and type of interceptor used shall be determined by a code official.

True False

3) Sand interceptors must have a water seal of not less than 3 inches.

True False

4) Interceptors must be vented.

True False

5) Interceptors must not be allowed to become air bound.

True False

6) The cover of any interceptor must be readily accessible.

True False

7) The water connection for cooling or operating an interceptor must be protected from back flow.

True False

8) Interceptors used in commercial laundries must prevent strings, rags, buttons, and other detrimental materials from entering the drainage system.

True False

9) Backwater valves are not required to ensure a mechanical seal against back flow.

True False

10) Backwater valves shall be installed in the branch of a building drain that receives only the discharge from fixtures located within such branch and shall be located above grade.

True False

True-false test for zone 3

1) Backwater valves must be installed so as to be readily accessible.

True False

2) Interceptors must be installed so as to be readily accessible.

True False

3) Fixtures that are subject to back flow from a public sewer must be protected with the use of a backwater valve.

True False

4) Grease interceptors are required to receive the drainage from fixtures and equipment with grease-laden waste located in food preparation areas.

True False

5) The rate of water flow through grease interceptors need not be regulated in commercial establishments where the seating limitations are 25 people or less.

True False

6) Vitrified clay traps, when installed underground, must be embedded in concrete.

True False

7) Under normal conditions, building traps shall be installed, at the option of a master plumber, either inside or outside the foundation of a home.

True False

8) Fixture traps must be sized to drain a fixture at a slow, steady pace.

True False

9) A combination fixture is defined as a fixture combining one sink and laundry tray or a two- or three-compartment sink or laundry tray in one unit.

True False

10) Slip joints are permitted on the trap inlet, trap outlet, or within the trap seal.

True False

Congratulations; you've completed all the tests on traps, interceptors, separators, and backwater valves. Did you have any idea there was so much to learn about these items? How do you think you did? Well, turn to the answers at the end of the chapter and check your work. Don't feel too badly if you didn't score well, many plumbers and potential plumbers have trouble with these questions. After all, if the questions didn't challenge your knowledge, they wouldn't do much to prepare you for the licensing exam.

Once you have checked your answers and researched any answers that were incorrect, move on to chapter 11. There you will get an introduction to the world of storm water and the drainage systems designed to control it. Roof drains and related storm-water drainage can present some problems for inexperienced plumbers, so you won't want to rush through the chapter. Take your time checking your answers to the questions in this chapter, and then proceed to chapter 11.

Answers to multiple-choice exam

1) D	16) D	31) D	46) B
2) B	17) D	32) B	47) C
3) D	18) C	33) C	48) B
4) B	19) C	34) B	49) C
5) B	20) B	35) A	50) A
6) D	21) C	36) D	51) B
7) C	22) C	37) D	52) B
8) D	23) A	38) B	53) B
9) D	24) C	39) C	54) A
10) D	25) B	40) A	55) A
11) B	26) C	41) D	56) B
12) C	27) C	42) B	57) D
13) B	28) C	43) D	58) B
14) B	29) D	44) B	59) B
15) C	30) A	45) A	60) C

Answers to true-false test for zone 1

1) T	5) T	9) T
2) T	6) F	10) T
3) F	7) F	11) T
4) T	8) T	12) F

13) F	19) F	25) T
14) T	20) T	26) T
15) T	21) T	27) F
16) F	22) F	28) F
17) T	23) T	29) T
18) T	24) T	30) T

Answers to true-false test for zone 2

1) T	5) T	8) T
2) T	6) T	9) F
3) F	7) T	10) F
4) T		

Answers to true-false test for zone 3

1) F	5) F	8) F
2) T	6) T	9) T
3) T	7) F	10) T
4) T		

Table 10-1 **Trap Sizes for Zone 1**

Type of fixture	Trap size
Bathtub	1½"
Shower	2"
Residential toilet	Integral
Lavatory	1¼"
Bidet	1½"
Laundry tub	1½"
Washing-machine standpipe	2"

Floor drain	2"
Kitchen sink	1½"
Dishwasher	1½"
Drinking fountain	1¼"
Public toilet	Integral

Fixture-Unit Requirements on Trap Sizes for Zone 2

Table 10-2

1¼" trap	1 fixture-unit
1½" trap	2 fixture-units
2" trap	3 fixture-units
3" trap	5 fixture-units
4" trap	6 fixture-units

Trap Sizes for Zone 2

Table 10-3

Type of fixture	Trap size
Bathtub	1½"
Shower	2"
Residential toilet	Integral
Lavatory	1¼"
Bidet	1½"
Laundry tub	1½"
Washing-machine standpipe	2"
Floor drain	2"
Kitchen sink	1½"
Dishwasher	1½"
Drinking fountain	1"
Public toilet	Integral

Table 10-4

Trap Sizes for Zone 3

Type of fixture	Trap size
Bathtub	1½"
Shower	2"
Residential toilet	Integral
Lavatory	1¼"
Bidet	1¼"
Laundry tub	1½"
Washing-machine standpipe	2"
Floor drain	2"
Kitchen sink	1½"
Dishwasher	1½"
Drinking fountain	1¼"
Public toilet	Integral
Urinal	2"

CHAPTER 11

The basics of storm water drainage systems

THE basics of storm water drainage systems are easy to understand. These drainage systems are utilized to control the build-up and runoff of rainwater. This, of course, is oversimplifying the purpose of storm water drainage, but it is the most basic definition of the systems.

When it comes to computing and designing storm water systems, the process can get quite complicated. The math alone can be intimidating. However, as a potential master plumber, you will have to learn how to design storm water systems. This job is sometimes left to engineers and architects, but the licensing exam will test your knowledge of the subject. Refer to Tables 11-1 through 11-7 at the end of this chapter for approved materials by zone.

The installation of storm water drainage system is not the same as that of a system meant to transfer sewage and waste water, although many of the principles are the same. For example, roof drains are installed whereas sanitary drains are not. Some locations allow the commingling of storm water and waste water, but other jurisdictions prohibit the combined use of a sewer system.

When you talk about a sanitary drainage system, a vertical pipe that conveys waste from one story to another is called a stack. If you install a drainage system for storm water, a similar pipe is called a leader or a conductor, depending on where it is located.

Subsoil and area drains are examples of yet other different systems. Subsoil drains collect subsurface water and convey it to a point of disposal. Area drains are similar to a floor drain, although they are not the same thing. These drains are examples of how storm water differs from sanitary drainage.

If you were given a detailed set of blueprints for a commercial building and asked to label all of the drains as either sanitary drains or storm drains, the task might seem difficult if you don't understand the differences between the two types of drainage. What appears to be a floor drain in an open, outside stairway will actually be an area drain, not a floor drain. This type of confusion might cost you valuable points when the licensing exam is scored.

Licensing exams for plumbers vary in complexity and content. In fact, they are often changed from time to time to prevent cheating. Since I cannot predict with absolute certainty what will be on the licensing exam, I will test you on a majority of the material that you might encounter. In all probability, the tests that I give you in this chapter will be more difficult than what you will encounter on the licensing exam. Of course, there is no way of knowing exactly what you will face on the day of the test, so it is better to be overprepared than underprepared. With that said, let's get on with the multiple-choice exam.

Multiple-choice exam

1) A receptacle installed to collect and drain storm water from an open area is called:

   A) A floor drain
   B) An area drain
   C) A split drain
   D) A combination sewer

2) A building drain that conveys storm water but no sewage is called:

   A) A sanitary drain
   B) A surface drain
   C) A storm drain
   D) A combined drain

3) A drain that collects water from beneath the surface and conveys it to a point of disposal is called:

   A) A combined drain
   B) A gray-water drain
   C) An underground drain
   D) A subsoil drain

4) Which of the following is required in the installation of a storm drainage system:

A) 45-degree bends
B) 90-degree bends
C) Cleanouts
D) Blind plugs

5) Which of the following commonly carries both storm water and sewage:

A) A combined sewer
B) A sanitary sewer
C) A mixed drain
D) A subsoil drain

6) Which of the following is not allowed in the construction of a storm water drainage system:

A) Drainage fittings that are commonly used for sanitary drainage
B) Schedule-40 plastic fittings
C) Fittings that retard the flow of storm water
D) Cast-iron fittings

7) Exterior pipes used to convey storm water from a roof drain to a point of disposal are known as:

A) Stacks
B) Wet stacks
C) Leaders
D) Conductors

8) Interior pipes used to convey storm water from a roof drain to a point of disposal are known as:

A) Stacks
B) Wet stacks
C) Leaders
D) Conductors

9) Roof drains are designed to discharge storm water into:

A) Leaders
B) Conductors
C) Either A or B
D) None of the above

10) A combined sewer is one that conveys:

A) Storm water
B) Sewage
C) Waste
D) All of the above

11) Under normal conditions, horizontal drains for storm water must have a minimum pitch of _____ inch per foot. Fill in the blank.

A) 1/16
B) 1/8
C) 1/4
D) 1/2

12) When a building subdrain is installed below the level of a public sewer, the subdrain must discharge into:

A) A septic tank
B) Sump
C) Receiving tank
D) Either B or C

13) When storm drains are allowed to connect to a sanitary sewer, the connection must be made at least _____ feet downstream from any soil stack. Fill in the blank.

A) 4
B) 5
C) 10
D) 12

14) Control devices used with controlled-flow roof drains must be:

A) No less than 4 inches in diameter
B) Electronic
C) Protected by strainers
D) Used as cleanouts

15) Conductors shall not be used as:

A) Soil pipes
B) Waste pipes
C) Vent pipes
D) All of the above

16) When a leader is connected to a combined sewer, it must be:

A) At least 4 inches in diameter
B) Trapped
C) Readily accessible
D) All of the above

17) Sump pits may be constructed of:

A) Tile
B) Concrete
C) Steel
D) All of the above

18) Sump pits may be constructed of:

A) Plastic
B) Concrete
C) Steel
D) All of the above

19) Sumps for subsoil drains are not required to:

A) Have gastight covers
B) Be vented
C) Both A and B
D) None of the above

20) Sumps for subsoil drains must be emptied:

A) Weekly
B) Daily
C) Automatically
D) Manually

21) A sump-pump system consists of:

A) A sump pump
B) A sump pit
C) Discharge piping
D) All of the above

22) Sump pits must be fitted with:

A) Gastight covers
B) Removable covers
C) Hinged covers
D) Airtight covers

23) When a storm drain discharges into a sanitary sewer, it must be equipped with:

A) A free-flow check valve
B) An antisiphon device
C) Both A and B
D) Either A or B

24) When a subsoil drain is subject to back flow, it must be protected with:

A) A vacuum breaker
B) A gate valve
C) A backwater valve
D) Any of the above

25) Subsoil drains may discharge into:

A) A trapped area drain
B) A sump
C) Either A or B
D) None of the above

26) Subsoil drains may discharge into:

A) A dry well
B) A sump
C) Either A or B
D) None of the above

27) The size of a trap for an individual conductor is determined by the size of:

A) The roof being drained
B) The horizontal drain to which it is connected
C) The sanitary sewer
D) All of the above

28) When a backwater valve is installed, it must be provided with:

A) Access
B) A gate valve
C) A check valve
D) Both B and C

29) An acceptable method of protecting an exposed leader is the use of:

A) Highly visible paint
B) Caution signs
C) Metal guards
D) Any of the above

30) When roof drains with strainers are installed, they must extend _____ inches above the roof. Fill in the blank.

A) 2
B) 4
C) 6
D) 8

31) Roof drains are required on all roof areas that do not:

A) Drain into hanging gutters
B) Have a curb
C) Either A or B
D) None of the above

32) A roof drain must have an available inlet area that is not less than _____ times the area of the conductor or leader to which the drain is connected. Fill in the blank.

A) 2
B) 3
C) 4
D) 6

33) A controlled-flow system is designed based on:

A) The size of the sanitary sewer
B) The size of the storm sewer
C) The local rainfall rate
D) All of the above

34) A roof with an area of 9,000 square feet requires a minimum of:

A) 1 roof drain
B) 2 roof drains
C) 3 roof drains
D) 4 roof drains

35) A roof with an area of 19,000 square feet requires a minimum of:

A) 1 roof drain
B) 2 roof drains
C) 3 roof drains
D) 4 roof drains

36) The strainers installed on roof drains are allowed to be of a flat-surface type when they are installed on:

A) Parking decks
B) Sun decks
C) Either A or B
D) None of the above

37) When flat-surface type strainers are used on roof drains, they must have an available inlet area that is not less than ____ times the area of the conductor or leader that receives the drainage. Fill in the blank.

A) 2
B) 3
C) 4
D) 6

38) Discharge piping and fittings installed on sump-pump systems must be:

A) Nylon
B) Plastic
C) The same size, or larger than, the discharge tapping of the sump pump
D) The same size, or smaller than, the discharge tapping of the sump pump

39) The floor of a sump pit must provide:

A) A minimum area of 10 square feet
B) A minimum area of 8 square feet
C) A permanent support for a sump pump
D) Both B and C

40) When cleanouts are installed with storm-water systems, they must be provided with:

A) Raised heads
B) Recessed heads
C) Access
D) Identification tags

41) Which of the following is a suitable material for use with roof drains?

A) Copper
B) Cast iron
C) Both A and B
D) Neither A nor B

42) Which of the following is a suitable material for use with roof drains?

A) Lead
B) Cast iron
C) Both A and B
D) Neither A nor B

43) We are now going to talk about sizing requirements for storm water systems. We will start with vertical walls. As you read this question, and the next few that follow, you will have to prove your working knowledge of vertical walls in the calculation of roof drains and storm water systems. For example, the question here is this: If you have a vertical wall that rises above a roof in a way that water will hit the vertical wall and fall onto the roof, how will that affect the sizing of a storm water system? Assuming that you have one vertical wall, what percentage of the wall area must you add to the roof area calculations?

A) 20 percent
B) 25 percent
C) 35 percent
D) 50 percent

44) Based on the information provided in the last question, what percentage of the total vertical wall area would you have to add to the roof area calculations if you had two adjacent vertical walls?

A) 20 percent
B) 25 percent
C) 35 percent
D) 50 percent

45) Based on the information provided in question 43, what percentage of the total vertical wall area would you have to add to the roof area calculations if you had two vertical walls that were the same height and located opposite of each other?

A) 0 percent
B) 10 percent
C) 15 percent
D) 50 percent

46) Based on the information provided in question 43, what percentage of the total vertical wall area would you have to add to the roof area calculations if you had two vertical walls that were of different heights and located opposite of each other?

A) 0 percent
B) 10 percent of the wall area above the top of the lower wall
C) 15 percent of the wall area above the top of the lower wall
D) 50 percent of the wall area above the top of the lower wall

47) Based on the information provided in question 43, what percentage of the total vertical wall area would you have to add to the roof area calculations if you had four vertical walls that were the same height and located opposite of each other?

A) 0 percent
B) 10 percent
C) 15 percent
D) 50 percent

48) Based on the information provided in question 43, what percentage of the total vertical wall area would you have to add to the roof area calculations if you had three vertical walls that were the same height and located opposite of, or adjacent to, each other?

A) 0 percent
B) 10 percent
C) 15 percent
D) 50 percent

49) Based on a rainfall of 1 inch per hour, how may square feet of drainage area is a 2-inch drain or leader able to drain?

A) 1400
B) 2400
C) 2880
D) 2950

50) Based on a rainfall of 2 inches per hour, how may square feet of drainage area is a 2-inch drain or leader able to drain?

A) 1400
B) 1440
C) 2000
D) 2880

51) Based on a rainfall of 1 inch per hour, how may square feet of drainage area is a 3-inch drain or leader able to drain?

A) 6400
B) 7400
C) 8800
D) 9200

52) Based on a rainfall of 2 inches per hour, how may square feet of drainage area is a 3-inch drain or leader able to drain?

A) 4400
B) 4460
C) 4750
D) 5880

53) Based on a rainfall of 1 inch per hour, how may square feet of drainage area is a 4-inch drain or leader able to drain?

A) 1840
B) 18400
C) 9200
D) 19200

54) Based on a rainfall of 2 inches per hour, how may square feet of drainage area is a 4-inch drain or leader able to drain?

A) 7200
B) 8220
C) 8900
D) 9200

55) Based on a rainfall of 2 inches per hour, how many square feet of drainage area is a 3-inch, horizontal pipe able to drain if the pipe is installed with a pitch of ⅛ of an inch per foot?

A) 1444
B) 1544
C) 1644
D) 2244

56) Based on a rainfall of 2 inches per hour, how many square feet of drainage area is a 4-inch, horizontal pipe able to drain if the pipe is installed with a pitch of ⅛ of an inch per foot?

A) 2769
B) 3760
C) 3969
D) 5200

57) Based on a rainfall of 2 inches per hour, how many square feet of drainage area is a 3-inch, horizontal pipe able to drain if the pipe is installed with a pitch of ¼ of an inch per foot?

A) 1820
B) 2020
C) 2320
D) 2520

58) Based on a rainfall of 2 inches per hour, how many square feet of drainage area is a 4-inch, horizontal pipe able to drain if the pipe is installed with a pitch of ¼ of an inch per foot?

A) 4300
B) 5300
C) 6300
D) 6750

59) The strainers installed on roof drains must extend a minimum of _____ inches above the surface of the roof. Fill in the blank.

A) 4
B) 5
C) 6
D) 8

60) The minimum inlet area of a strainer for a roof drain must be at least _____ times the size of the pipe it connects to. Fill in the blank.

A) 1
B) 1½
C) 2
D) 2½

Now that you've completed the multiple-choice exam, how do you feel about your knowledge of storm water drainage? Unless you are responsible for sizing a storm water system, the rules and regulations that pertain to them are not too difficult to grasp. Let's move now to the true-false test for your region and continue our examination of your knowledge.

True-false test for zone 1

1) Piping installed on the exterior of a building and used to convey rainwater must not be less than 26 gauge galvanized sheet metal.

   True False

2) Strainers for roof drains must have a minimum inlet area equal to 1½ times the size of the pipe the roof drain is connected to.

   True False

3) Flat-surface type roof drains must have a minimum inlet area equal to at least 2 times the size of the pipe the roof drain is connected to.

   True False

4) Roof drains may not be made from lead.

   True False

5) Roof drains may be made of cast iron or copper.

   True False

6) Rainwater systems installed within buildings are required to be tested with the same methods used to test drain, waste, and vent systems.

   True False

7) Roof drains are only required to be equipped with strainers when the roof is situated under overhanging trees.

   True False

8) Rainwater piping can be used as a wet vent for a sanitary drainage system.

True False

9) The use of ABS, PVC, and DWV pipe for rainwater systems is limited to buildings that have no more than 2 stories above ground.

True False

10) Soil pipes cannot be used as rainwater pipes.

True False

True-false test for zone 2

1) Expansion joints may be required on storm water piping.

True False

2) Pipes carrying storm water to a storm sewer are not required to be trapped.

True False

3) Pipes carrying storm water to a storm sewer are required to be vented.

True False

4) A sump-pump installation requires the use of a check valve.

True False

5) A sump-pump installation requires the use of a gate valve.

True False

6) Aluminum is a suitable material for the construction of a building storm sewer.

True False

7) Storm water piping is not required to be trapped when discharging into a sanitary sewer.

True False

8) Horizontally split pipe is prohibited from use as a subsoil drain material.

True False

9) Subsoil drains may not be placed under basement floors.

True False

10) Courtyards must be drained into a storm sewer system when such a system is available.

True False

True-false test for zone 3

1) In most cases, a sanitary sewer and a storm sewer can be combined in a common pipe.

True False

2) Double connections of storm drains are prohibited.

True False

3) Runoff control, in a controlled-flow roof drainage system, must be accomplished with the use of control devices.

True False

4) Leaders installed along alleyways must be protected from damage.

True False

5) Subsoil drains may not be discharged to locations above grade.

True False

6) The rules governing the placement of cleanouts in sanitary drainage systems apply to storm water systems.

True False

7) Storm water from a 2-family dwelling is normally permitted to discharge onto flat areas, such as streets and lawns.

True False

8) A combined building drain is a drain that conveys both storm water and sewage.

True False

9) The size of a drainage pipe may not be reduced in the direction of flow.

True False

10) The minimum pitch requirement for the installation of storm water drainage pipes is ¼ inch per foot.

True False

This brings us to the close of another chapter. Take some time to check your answers and research your mistakes. Spend enough time in doing this to gain a complete understanding of how the code relates to storm water drainage. When you feel this chapter is behind you, move on to chapter 12 and study the rules and regulations that pertain to vents and venting.

Answers to multiple-choice exam

1) B	16) B	31) A	46) D
2) C	17) D	32) A	47) A
3) D	18) D	33) C	48) D
4) C	19) C	34) B	49) C
5) A	20) C	35) D	50) B
6) C	21) D	36) C	51) C
7) C	22) B	37) A	52) A
8) D	23) D	38) C	53) B
9) C	24) C	39) C	54) D
10) D	25) C	40) C	55) C
11) B	26) C	41) C	56) B
12) D	27) B	42) C	57) C
13) C	28) A	43) D	58) B
14) C	29) C	44) C	59) A
15) D	30) B	45) A	60) B

Answers to true-false test for zone 1

1) T
2) T
3) T
4) F
5) T
6) T
7) F
8) F
9) F
10) T

Answers to true-false test for zone 2

1) T
2) T
3) F
4) T
5) T
6) T
7) F
8) F
9) F
10) T

Answers to true-false test for zone 3

1) F
2) F
3) T
4) T
5) F
6) T
7) T
8) T
9) T
10) F

Table 11-1

Horizontal Storm Water Sizing Table

Pipe grade per foot	Pipe size	Gallons per minute (gpm)	Number of square feet of surface area
¼"	3"	48	4,640
¼"	4"	110	10,600
¼"	6"	314	18,880
¼"	8"	677	65,200

Table 11-2 **Approved Materials for Storm Water Drainage in Zone 1**

Materials approved for use inside buildings, above ground
Galvanized materials
Wrought-iron materials
Brass materials
Copper materials
Cast-iron materials
ABS materials*
PVC materials*
Lead materials

*Note—ABS and PVC may not be used in buildings that have more than three floors above ground.

Table 11-3 **Approved Materials for Exterior Storm Water Drainage in Zone 1**

Materials approved for use on the exterior of buildings
Sheet metal with a minimum gauge of 26

Table 11-4 **Approved Materials for Storm Water Drainage in Zone 1**

Materials approved for use inside buildings, below ground
Service-weight cast-iron materials
DWV copper materials
ABS materials
PVC materials
Extra-strength vitrified-clay materials

Approved Materials for Storm Water Drainage in Zone 2 — Table 11-5

Materials approved for above-ground use

Galvanized materials

Black-steel materials

Brass materials

DWV copper materials or thicker types of copper

Cast-iron materials

ABS materials

PVC materials

Aluminum materials

Lead materials

Approved Materials for Storm Water Drainage in Zone 2 — Table 11-6

Materials approved for underground use

Cast-iron materials

Coated aluminum materials

ABS materials*

PVC materials*

Copper materials*

Concrete materials*

Asbestos-cement materials*

Vitrified-clay materials*

*Note—These materials might be allowed for use, subject to local code authorities.

Table 11-7 **Approved Materials for Storm Water Drainage in Zone 2**

Materials approved for building storm sewers

Cast-iron materials

Aluminum materials*

ABS materials

PVC materials

Vitrified-clay materials

Concrete materials

Asbestos-cement materials

*Note—Buried aluminum must be coated.

CHAPTER 12

Installing vents and venting

THE installation of vents can get complicated. Without proper venting, a drainage system will not work well. Even if you didn't have to worry about plumbing inspectors rejecting your work, you will have dissatisfied customers if you fail to vent plumbing fixtures. Many people, licensed plumbers included, don't understand fully the primary purpose of vents. Do you know the main reason why vents are installed for each plumbing fixture?

If you asked 100 people why plumbing fixtures are vented, a large percentage of them would agree that vents are installed to keep sewer gas out of buildings. They think that without the vents, sewer gas would enter the building and become a problem. Would you agree with this group of people?

Vents in a plumbing system are installed to protect trap seals. It is the fixture's trap that prevents sewer gas from entering a building. While a majority of people think traps are installed only to prevent foreign objects from entering the drainage pipes, the traps are actually there to keep odors and dangerous gases from entering habitable space. However, without the proper venting, a trap seal might be siphoned down the drain and leave the building unprotected from sewer gas and odors.

Vents provide air circulation to the drainage system. They allow the drains to flow faster, and they prevent siphonic action in the traps. These are the two main purposes of vents in a plumbing system. Fixtures that are not vented adequately drain slowly. This same principle can be demonstrated with a can of juice.

Have you ever punched a hole in the top of a can of juice and tried to pour the contents from the can? If you have, you know that the liquid comes out slowly and in a jerky motion, sort of a glug-glug-glug motion. But, if you punch a second hole in the top of the can, opposite the location of the first hole, the liquid pours out faster and smoother. This is a fine example of how a plumbing vent improves the flow of drainage. The second hole made in the top of the can acts as a vent for the hole from which the liquid is poured. The same basic principle applies to a plumbing fixture. Without a vent, a fixture can't drain as well as it will with a vent. The key is air circulation.

Not all plumbing vents are vertical pipes that extend to open air. Some vents are wet vents, which means that the vent for one fixture serves as a drain for another fixture. In combination waste-and-vent systems, the drains are oversized to allow air circulation above the flow of drainage. There are many options for venting a plumbing system, and if you want to pass the licensing exam, you had better know most of them. Refer to Tables 12-1 through 12-8 at the end of this chapter for vent sizing and trap-to-vent distances by zone.

Now, let's discuss the issue of protecting trap seals. When a trap is not properly vented, there is a risk that the water that forms the trap seal will be sucked down the drain and leave the trap without enough water to block odors and gases. Most people have siphoned some type of liquid at some time in their lives. As you probably know, once a siphon hose has been started, the flow in the hose will continue until air is introduced into the hose or until the hose is raised above the level of the liquid being siphoned. This same principle applies to a drainage system.

Plumbing drains are typically installed at levels lower than the fixtures they serve. This is always the case, unless a pump is used to empty the fixture or sump. This means that the drain will not be raised above the level of the fixture, so siphonic action can't be broken that way.

If you were to fill a bathtub with water and then open the drain, the water would pass through the trap and down the drain. Without a vent, the volume of water draining down the pipe could create a siphonic action and pull all water from above the drain into the drainage system, including the water in the trap. Should this happen, the fixture would be left unprotected from odors and gases.

The trap might be replenished when a small quantity of water is run through it, but the next time a large volume of water is drained, the trap could be emptied again. This, of course, is not an acceptable risk. Therefore, the trap should be vented to avoid the problem. The vent allows air circulation in the drainage system, which prevents the possibility of siphonic action in the drain.

Now you know more about the true purpose of vents than some master plumbers. I have known many plumbers throughout my career who didn't understand why vents were used. They installed them because code requirements dictated the need for vents, but the plumbers couldn't explain why the vents were needed. You are about to be tested on the proper installation of vents, and I thought the knowledge might be easier to gain if you had a good understanding of why vents are needed and how they work. Now, let's get on with the multiple-choice exam.

Multiple-choice exam

1) Which of the following types of vent systems must be independent of a sanitary vent system?

A) Wet-vent system
B) Dry-vent system
C) Chemical-vent system
D) None of the above

2) A vent that connects to one or more individual vents with a stack vent or a vent stack is known as:

A) A yoke vent
B) A wet vent
C) A main vent
D) A branch vent

3) The primary vent in a venting system to which various vent branches are connected is called a:

A) Main vent
B) Stack vent
C) Vent stack
D) Yoke vent

4) A vent that serves only one fixture drain is called:

A) A fixture branch
B) A circuit vent
C) An individual vent
D) A common vent

5) When a soil or waste stack is extended above the highest horizontal drain connected to the stack, which of the following is created:

A) A code violation
B) A vent stack
C) A stack vent
D) A relief vent

6) Vents that connect to horizontal drainage branches and vent more than one trap are called:

A) Fixture branches
B) Circuit vents
C) Vent stacks
D) Relief vents

7) Vents are intended to protect:

A) Traps
B) Trap seals
C) Fixture outlets
D) All of the above

8) Vents used to protect against pressure changes in a stack are called:

A) Pneumatic vents
B) Wet vents
C) Circuit vents
D) Yoke vents

9) A vertical pipe installed to provide air circulation to and from the drainage system is called:

A) A stack vent
B) A vent stack
C) Leader
D) Conductor

10) An auxiliary vent that provides additional circulation of air in or between a drainage and vent system is called:

A) A branch vent
B) An island vent
C) A relief vent
D) A common vent

11) The section of pipe in a drainage system that runs from the trap of a fixture to a junction with any other drain pipe is known as:

A) A horizontal branch
B) A fixture drain
C) A fixture branch
D) None of the above

12) Drains that serve one or more fixtures that discharge into another drain or into a stack are called:

A) Wet vents
B) Fixture drains
C) Fixture branches
D) None of the above

13) Vents that serve the containers of sewage ejectors, or similar equipment, and terminate separately to open air are called:

A) Yoke vents
B) Pit vents
C) Sump vents
D) Pump vents

14) The minimum size of any vent pipe shall not be less than:

A) 1 inch
B) 1¼ inches
C) 1½ inches
D) 2 inches

15) The size of any vent with a developed length of more than 40 feet must be increased by:

A) 1 inch
B) 1 pipe size
C) 50 percent
D) None of the above

16) When determining the developed length of a vent, one must measure from the farthest point of the vent connection to the drainage system to the point of:

A) Connection to a vent stack
B) Connection to a stack vent
C) Termination into open air
D) Any of the above

17) The diameter of vents must be, at a minimum, equal to:

A) The size of the drain being served
B) One-half the size of the drain being served
C) Twice the size of the drain being served
D) None of the above

18) Individual vents must be connected to the _____ of the trap or trapped fixture being served. Fill in the blank.

A) Top
B) Bottom
C) Fixture drain
D) Fixture branch

19) A vent system for chemical waste must:

A) Not be run horizontally
B) Connect to a vent stack
C) Connect to a stack vent
D) Terminate separately through the roof to open air

20) A stack vent is the same as:

A) A vent stack
B) An individual vent
C) A branch vent
D) None of the above

21) The length of the pipeline that is measured along the centerline of the pipe and fittings is the:

A) Combined length
B) Determined length
C) Developed length
D) Maximum length

22) Vents must protect trap seals in a way that the trap seal is not subjected to a pneumatic pressure differential equal to, or exceeding:

A) A 1-inch water column
B) A 1½-inch water column
C) A 2-inch water column
D) None of the above

23) All sanitary drainage systems that receive the discharge of a water closet are required to have:

A) A vent stack
B) A stack vent
C) Either A or B
D) Both A and B

24) Every vent stack must terminate:

A) Through a roof
B) Outdoors
C) Near a window
D) Both A and B

25) When a drainage stack consists of 5 branch intervals or more, a _____ is required. Fill in the blank.

A) Vent stack
B) Stack vent
C) Combined vent
D) Relief vent

26) All vent stacks must connect to:

A) The base of a drainage stack
B) The base of a sewer
C) The base of a fixture
D) Any of the above

27) When a vent stack is connected to a building drain, the connection must be made within _____ pipe diameters downstream of the drainage stack. Fill in the blank.

A) 2
B) 4
C) 5
D) 10

28) If there are 5 or more branch intervals located above a horizontal offset of a drainage stack, the offsets must be:

A) No more than 22½ degrees
B) Trapped
C) Vented
D) Eliminated

29) The length of a vent terminal when it penetrates a roof must be no less than _____ inches above the roof. Fill in the blank.

A) 4
B) 6
C) 8
D) 12

30) When a roof is used for some purpose other than weather protection, the extension of a vent penetrating the roof must not be less than:

A) 6 inches
B) 12 inches
C) 4 feet
D) 7 feet

31) Vent terminals must not be installed directly beneath any:

A) Window
B) Door
C) Ventilating opening
D) All of the above

32) When a vent pipe is installed on the outside of a building where freezing temperatures might occur, the vent must be protected from freezing by which of the following means:

A) Insulation
B) Heat
C) Either A or B
D) Both A and B

33) Vent terminals may not be installed:

A) In attics
B) Through walls
C) Through roofs
D) All of the above

34) A vent may not terminate within _____ feet of a window, unless it is extended to a point at least _____ feet above the window. Fill in the blanks.

A) 5 and 2
B) 2 and 5
C) 10 and 2
D) 10 and 5

35) Vents that terminate through a wall, must be at least _____ feet from the lot line of the property. Fill in the blank.

A) 5
B) 10
C) 15
D) 20

36) Vents that terminate through a wall, must be at least _____ feet above the ground. Fill in the blank.

A) 5
B) 10
C) 15
D) 20

37) Vent terminals may not terminate under the overhang of the building when the overhang:

A) Extends less than 4 inches
B) Contains soffit vents
C) Will not protect the opening of the vent
D) All of the above

38) Side-wall vents must be protected to prevent birds and rodents from:

A) Entering the vent opening
B) Blocking the vent opening
C) Both A and B
D) Either A or B

39) With one exception, all individual, branch, and circuit vents must:

A) Connect to a vent stack
B) Connect to a stack vent
C) Terminate into open air
D) Any of the above

40) All dry vents must rise vertically to a point at least _____ inches above the flood-level rim of the highest trap or trapped fixture being vented. Fill in the blank.

A) 4
B) 6
C) 12
D) 42

41) All vents should be graded and connected in a manner that will allow them to:

A) Drain back to a soil or waste pipe by gravity
B) Maintain a grade of ⅛ of an inch to the foot
C) Be rodded out from the roof
D) Maintain a grade of ½ of an inch to the foot

42) Which of the following types of vents must, when connected to a horizontal drain, connect above the centerline of the horizontal drain pipe:

A) Dry vents
B) Wet vents
C) Combination vents
D) All of the above

43) Horizontal vent pipes that form circuit vents must be connected at a point at least _____ inches above the flood-level rim of the highest fixture being vented. Fill in the blank.

A) 4
B) 6
C) 8
D) 42

44) Horizontal vent pipes that form branch vents must be connected at a point at least _____ inches above the flood-level rim of the highest fixture being vented. Fill in the blank.

A) 4
B) 6
C) 8
D) 42

45) Horizontal vent pipes that form relief vents must be connected at a point at least _____ inches above the flood-level rim of the highest fixture being vented. Fill in the blank.

A) 4
B) 6
C) 8
D) 42

46) When a vent pipe is connected to a vent stack or a stack vent, the connection must be made at least _____ inches above the flood-level rim of the highest fixture being vented. Fill in the blank.

A) 4
B) 6
C) 8
D) 42

47) Vents that are roughed-in for future fixtures must:

A) Connect to the vent system
B) Be labeled as a vent
C) Both A and B
D) None of the above

48) Any combination of fixtures within _____ bathroom groups located on the same floor level may be wet vented. Fill in the blank.

A) 2
B) 3
C) 4
D) Any of the above

49) A dry vent connection to a wet vent must be an individual vent or a common vent to which of the following:

A) Bidet
B) Lavatory
C) Bathtub
D) Any of the above

50) A dry vent connection to a wet vent must be an individual vent or a common vent to which of the following:

A) Shower
B) Lavatory
C) Bathtub
D) Any of the above

51) A wet vent with a diameter of 1½ inches may accept the discharge of how many fixture units?

A) 1
B) 2
C) 4
D) 6

52) A wet vent with a diameter of 2 inches may accept the discharge of how many fixture units?

A) 1
B) 2
C) 4
D) 6

53) A wet vent with a diameter of 3 inches may accept the discharge of how many fixture units?

A) 3
B) 4
C) 6
D) 12

54) Individual vents may be used as a common vent to vent _____ traps or trapped fixtures, with one exception, that are located on the same floor level. Fill in the blank.

A) 2 C) 4
B) 3 D) 5

55) The exception to the rule in question 54 allows an individual vent to be used as a common vent for _____ lavatories when the fixtures are installed on the same floor level. Fill in the blank.

A) 3 C) 5
B) 4 D) 6

56) Fixtures being served by a common vent and connected at the same level shall connect to the vent at the interconnection of the:

A) Traps C) Fixture drains
B) Fixture branches D) Stack vent

57) A common vent with a diameter of 1½ inches may accept the maximum discharge from upper fixture drains in the amount of how many fixture units?

A) 1 C) 3
B) 2 D) 4

58) A common vent with a diameter of 2 inches may accept the maximum discharge from upper fixture drains in the amount of how many fixture units?

A) 2 C) 6
B) 4 D) 8

59) A common vent with a diameter of 3 inches may accept the maximum discharge from upper fixture drains in the amount of how many fixture units?

A) 2 C) 6
B) 4 D) 8

60) What is the maximum number of fixtures that may be connected to a horizontal branch drain and be vented with a circuit vent?

A) 3 C) 6
B) 4 D) 8

61) When a circuit vent is used, each fixture drain shall connect _____ to the horizontal branch being circuit vented. Fill in the blank.

A) Vertically
B) Horizontally
C) Indirectly
D) None of the above

62) Circuit vents are not allowed to:

A) Run for more than 10 feet
B) Receive the discharge from any waste pipe
C) Run horizontally
D) Connect to a stack vent

63) The maximum slope allowed on a vent section of a horizontal branch drain is:

A) ½ inch per foot
B) ¾ inch per foot
C) 1 inch per foot
D) 1¼ inches per foot

64) When a circuit-vented horizontal branch receives the discharge of _____ or more water closets, a relief vent must be provided. Fill in the blank.

A) 2
B) 3
C) 4
D) 6

65) Crown venting is not allowed within _____ pipe diameters of the trap weir. Fill in the blank.

A) 1½
B) 2
C) 2½
D) 3

66) Vents for fixture drains, excluding fixtures with integral traps, must connect above:

A) The flood-level rim of the fixture
B) The weir of the trap being vented
C) The fixture drain
D) None of the above

67) To determine the distance of a fixture trap from its vent, a plumber will normally consult a table in a code book. These tables, however, may not always be available, such as on the day of the licensing exam. To determine the distance limitation between a

trap and its fixture, you must know the size of the trap, the size of the fixture drain, and the slope of the pipe. With that information provided, you should be able to decide how far a trap can be placed from its vent. This question, and the next several questions that follow will test your knowledge on this issue. I will provide the information needed to make a correct decision, but you must know the maximum distances allowed by the plumbing code. Here is your first scenario: You have a fixture with a 1½-inch trap and drain. The slope of the pipe is ¼ inch per foot. How far from the vent may the trap for the fixture be placed?

A) 3½ feet
B) 4 feet
C) 5 feet
D) 6 feet

68) You have a fixture with a 1¼-inch trap and drain. The slope of the pipe is ¼ inch per foot. How far from the vent may the trap for the fixture be placed?

A) 3½ feet
B) 4 feet
C) 5 feet
D) 6 feet

69) You have a fixture with a 1¼-inch trap and a 1½-inch drain. The slope of the pipe is ¼ inch per foot. How far from the vent may the trap for the fixture be placed?

A) 3½ feet
B) 4 feet
C) 5 feet
D) 6 feet

70) You have a fixture with a 1½-inch trap and a 2-inch drain. The slope of the pipe is ¼ inch per foot. How far from the vent may the trap for the fixture be placed?

A) 42 inches
B) 4 feet
C) 6 feet
D) 8 feet

71) You have a fixture with a 2-inch trap and a 2-inch drain. The slope of the pipe is ¼ inch per foot. How far from the vent may the trap for the fixture be placed?

A) 42 inches
B) 4 feet
C) 6 feet
D) 8 feet

72) You have a fixture with a 3-inch trap and a 3-inch drain. The slope of the pipe is ⅛ inch per foot. How far from the vent may the trap for the fixture be placed?

A) 6 feet
B) 8 feet
C) 10 feet
D) 12 feet

73) You have a fixture with a 4-inch trap and a 4-inch drain. The slope of the pipe is ⅛ inch per foot. How far from the vent may the trap for the fixture be placed?

A) 6 feet
B) 8 feet
C) 10 feet
D) 12 feet

74) A relief vent must be provided for every drainage stack that has more than _____ branch intervals. Fill in the blank.

A) 4
B) 5
C) 8
D) 10

75) When a relief vent is provided for a drainage stack, the relief vent must:

A) Connect to a stack vent
B) Connect to a vent stack
C) Either A or B
D) Be installed horizontally

76) Which of the following fixtures may be served by a combination waste-and-vent system?

A) Floor drains
B) Sinks
C) Lavatories
D) Any of the above

77) Which of the following fixtures may not be served by a combination waste-and-vent system?

A) Standpipes
B) Water closets
C) Lavatories
D) A and B

78) The maximum vertical distance for a vertical pipe in a combination waste-and-vent system is:

A) 30 inches
B) 4 feet
C) 8 feet
D) 10 feet

79) The maximum slope allowed on a horizontal combination waste-and-vent pipe is:

A) ¼ inch per foot
B) ½ inch per foot
C) ¾ inch per foot
D) 1 inch per foot

80) The only vertical pipes allowed in a combination waste-and-vent system are the connections between horizontal pipes and the fixture branches of:

A) Lavatories
B) Sinks
C) Floor drains and standpipes
D) All of the above

81) When there is a danger of frost closure in a vent, as determined by the plumbing code, the vent terminal must be a minimum of _____ inches in diameter, and if the vent size must be increased to accomplish this size, the increase shall take place at least _____ inches below the roof or inside wall where the vent penetrates. Fill in the blanks.

A) 1½ and 6
B) 2 and 12
C) 3 and 12
D) 4 and 12

82) A waste stack vent must be installed:

A) Without offsets
B) In conjunction with a chemical stack
C) Vertically
D) In buildings with more than 2 stories above ground

83) When a stack vent is provided for a waste stack, the stack vent must be sized to be:

A) No more than 1 pipe diameter smaller than the waste stack
B) The same size as the waste stack
C) 1 pipe size larger than the waste stack
D) No less than 4 inches in diameter

84) Island fixture venting is allowed for which of the following fixtures:

A) Sinks
B) Bidets
C) Floor drains
D) All of the above

85) Island fixture venting is allowed for which of the following fixtures:

A) Sinks
B) Bidets
C) Lavatories
D) Either A or C

86) Island fixture venting is allowed for which of the following fixtures:

A) Residential kitchen sinks
B) Residential kitchen sinks with dishwasher waste connections
C) Residential kitchen sinks with garbage disposers
D) Any of the above

87) Island vents must rise vertically to a point:

A) Above the flood-level rim of the fixture being vented
B) At least 6 inches above the flood-level rim of the fixture being vented
C) Above the drainage outlet of the fixture being vented
D) At least 12 inches above the base of the cabinet containing the fixture

88) When island vents are used, _____ must be installed and accessible.

A) Cleanouts
B) Trap adapters
C) Sanitary tees
D) None of the above

89) The total discharge, in terms of fixture units, for a 2-inch waste stack is:

A) 1 fixture unit
B) 2 fixture units
C) 4 fixture units
D) 6 fixture units

This concludes the multiple-choice test on vents and venting. Depending on where you will be working and taking the licensing exam, there may be some differences between your local code and the questions provided in this exam. This same situation might apply to other multiple-choice exams in this book. Always confirm your answers with a current, local code reference.

Now we will move on to the true-false tests for the various regions. Since the true-false questions pertain only to one region, there is less chance of a discrepancy in the questions asked, but again, always confirm your answers with your own local plumbing code.

True-false test for zone 1

1) Every plumbing fixture, except as otherwise provided in the code, must be vented.

True False

2) The use of ABS and PVC piping is limited to use in buildings not having more than 3 stories above ground.

True False

3) A vent may exceed ⅓ of the maximum horizontal length allowed by the code if the size of the vent is increased by 1 pipe size for its entire, developed length.

True False

4) All vent pipes must have a minimum slope of ¼ inch per foot.

True False

5) Every vent must rise a minimum of 42 inches above the flood-level rim of the fixture being served by the vent.

True False

6) No vent may terminate above a roof with an exposed distance of less than 12 inches.

True False

7) All vents that terminate through a roof must extend at least 6 inches above the roof.

True False

8) Any vent terminating above a roof must be at least 12 inches away from any vertical surface.

True False

9) A vent installed within 10 feet of a roof used for purposes other than weather protection must extend at least 7 feet above the roof.

True False

10) The minimum diameter for a vent that terminates above a roof in an area where frost closure might be a problem is 3 inches.

True False

11) When a vent is installed in an area that might be affected by frost closure, the vent must extend at least 12 inches above the roof.

True False

12) Drainage stacks that extend for 10 stories or more must be installed in conjunction with a vent stack.

True False

13) Vent stacks, when installed in conjunction with drainage stacks in buildings with 10 or more stories must be connected to the drainage stack at every 5-story interval with a relief yoke vent.

True False

14) Yoke vents used to connect drainage stacks and vent stacks must be sized so that they are no more than 1 pipe diameter smaller than the stacks that they connect.

True False

15) Wet venting is limited to horizontal drainage piping.

True False

16) Vertical piping between any 2 consecutive inlet levels is considered to be a wet-vented section.

True False

17) Island vents must extend to a vertical height of not less than that of the drainboard for the fixture being served.

True False

18) Combination waste-and-vent systems are prohibited.

True False

19) When a combination waste-and-vent system is used, the drain pipes must be at least 2 pipe sizes larger than they would have to be in a drainage system installed with normal venting.

True False

20) Combination waste-and-vent systems are prohibited, except where structural conditions preclude the use of a conventional venting system.

True False

True-false test for zone 2

1) Vent extensions above a roof must rise vertically for at least 6 inches.

True False

2) Vent extensions above roofs being used for purposes other than weather protection must extend at least 7 feet above the roof.

True False

3) Vent pipes may not be used to support flag poles or similar items.

True False

4) Every building in which plumbing is installed, with one exception, shall be provided with at least one 4-inch vent.

True False

5) A garage that contains a clothes washer or laundry tub is allowed to have a main vent with a diameter of 1½ inches.

True False

6) Combination waste-and-vent systems can be used to collect the discharge of all types of fixtures, except water closets and urinals.

True False

7) A single bathroom group consists of a water closet, basin, shower or tub.

True False

8) One sink and one lavatory or one tub, or one shower, or three lavatories within 8 feet of developed length of a main-vented line may be installed on a 2-inch horizontal waste branch.

True False

9) Two lavatories within 5 feet of developed length of a main-vented line may be installed on a 1½-inch horizontal waste branch.

True False

10) Up to 3 fixture units may be placed on a soil stack on floors above the highest water closet or bathtub connection with the use of re-vents provided the soil stack has a minimum diameter of 2 inches.

True False

True-false test for zone 3

1) A pipe that connects upward from a soil or waste stack to a vent stack for the purpose of preventing pressure changes in the stack is a yoke vent.

True False

2) A vent stack is required for every drainage stack that is a minimum of 4 branch intervals or more.

True False

3) The size and length of a sump vent is determined by the diameter of the vent pipe and the discharge capacity of the pump that empties the sump.

True False

4) The abbreviation DFU stands for the words Drainage Fixture Units.

True False

5) A 3-inch wet vent can carry a fixture unit load of 12.

True False

6) Vents may not be terminated out of a side wall of a building.

True False

7) When a vent terminates within less than 10 feet of a window, the vent must be extended 2 feet above the window.

True False

8) Horizontal vent pipes forming a circuit vent must be located at least 8 inches above the flood-level rim of the highest fixture being served by the vent.

True False

9) Up to 3 lavatories may be vented by a common, individual vent.

True False

10) Circuit-vented horizontal branch drains are permitted to be connected together.

True False

Well, this has been a long chapter, but we have finally finished it. Check your answers against the correct answers at the end of the chapter, and then move on to the next chapter where we will study potable water systems.

Answers to multiple-choice exam

1) C	12) C	23) C	34) C
2) D	13) C	24) B	35) B
3) A	14) B	25) B	36) B
4) C	15) B	26) A	37) B
5) C	16) D	27) D	38) C
6) B	17) B	28) C	39) D
7) B	18) C	29) D	40) B
8) D	19) D	30) D	41) A
9) B	20) D	31) D	42) A
10) C	21) C	32) C	43) B
11) B	22) A	33) A	44) B

45) B
46) B
47) C
48) A
49) D
50) D
51) A
52) C
53) D
54) A
55) A
56) C
57) A
58) B
59) C
60) D
61) B
62) B
63) C
64) C
65) B
66) B
67) C
68) A
69) C
70) D
71) C
72) C
73) D
74) D
75) B
76) D
77) B
78) C
79) B
80) D
81) C
82) C
83) B
84) A
85) D
86) D
87) C
88) A
89) C

Answers to true-false test for zone 1

1) T
2) T
3) T
4) F
5) F
6) F
7) T
8) T
9) T
10) F
11) F
12) T
13) T
14) F
15) F
16) T
17) T
18) F
19) T
20) T

Answers to true-false test for zone 2

1) T
2) F
3) T
4) F
5) T
6) T
7) T
8) T
9) T
10) F

Answers to true-false test for zone 3

1) T	5) T	8) F
2) F	6) F	9) T
3) T	7) T	10) T
4) T		

Table 12-1

Trap-to-Vent Distances in Zone 1

Grade on drain pipe	Size of trap arm	Maximum distance between trap and vent
¼"	1¼"	2' 6"
¼"	1½"	3' 6"
¼"	2"	5'
¼"	3"	6'
¼"	4" and larger	10'

Table 12-2

Trap-to-Vent Distances in Zone 2

Grade on drain pipe	Fixture drain size	Trap size	Maximum distance between trap and vent
¼"	1¼"	1¼"	3' 6"
¼"	1½"	1¼"	5'
¼"	1½"	1½"	5'
¼"	2"	1½"	8'
¼"	2"	2"	6'
⅛"	3"	3"	10'
⅛"	4"	4"	12'

Table 12-3 **Sizing a Vent Stack for Wet Venting in Zone 2**

Wet-vented fixtures	Vent-stack size requirements
1–2 bathtubs or showers	2"
3–5 bathtubs or showers	2½"
6–9 bathtubs or showers	3"
10–16 bathtubs or showers	4"

Table 12-4 **Stack-Venting without Individual Vents in Zone 2**

Fixtures allowed to be stack-vented without individual vents*
Water closets
Basins
Bathtubs
Showers
Kitchen sinks, with or without dishwasher and garbage disposer

*Note—restrictions apply to this type of installation.

Table 12-5 **Sizing a Wet Stack Vent in Zone 2**

Pipe size of stack	Fixture-unit load on stack	Maximum length of stack
2"	4	30
3"	24	50
4"	50	100
6"	100	300

Table 12-6

Trap-to-Vent Distances in Zone 3

Grade on drain pipe	Fixture drain size	Trap size	Maximum distance between trap and vent
¼"	1¼"	1¼"	3' 6"
¼"	1½"	1¼"	5'
¼"	1½"	1½"	5'
¼"	2"	1½"	8'
¼"	2"	2"	6'
⅛"	3"	3"	10'
⅛"	4"	4"	12'

Table 12-7

Vent Sizing Table for Zone 3 (For use with vent stacks and stack vents)

Drain pipe size	Fixture-unit load on drain pipe	Vent pipe size	Maximum developed length of vent pipe
1½"	8	1¼"	50'
1½"	8	1½"	150'
1½"	10	1¼"	30'
1½"	10	1½"	100'
2"	12	1½"	75'
2"	12	2"	200'
2"	20	1½"	50'
2"	20	2"	150'
3"	10	1½"	42'
3"	10	2"	150'
3"	10	3"	1,040'
3"	21	1½"	32'
3"	21	2"	110'
3"	21	3"	810'
3"	102	1½"	25'

Table 12-7 **Continued**

Drain pipe size	Fixture-unit load on drain pipe	Vent pipe size	Maximum developed length of vent pipe
3"	102	2"	86'
3"	102	3"	620'
4"	43	2"	35'
4"	43	3"	250'
4"	43	4"	980'
4"	540	2"	21'
4"	540	3"	150'
4"	540	4"	580'

Table 12-8 **Vent Sizing Table for Zone 3 (For use with individual, branch, and circuit vents for horizontal drain pipes)**

Drain pipe size	Drain pipe grade per foot	Vent pipe size	Maximum developed length of vent pipe
1½"	¼"	1¼"	Unlimited
1½"	¼"	1½"	Unlimited
2"	¼"	1¼"	290'
2"	¼"	1½"	Unlimited
3"	¼"	1½"	97'
3"	¼"	2"	420'
3"	¼"	3"	Unlimited
4"	¼"	2"	98'
4"	¼"	3"	Unlimited
4"	¼"	4"	Unlimited

CHAPTER 13

Approved potable water supply systems

WHEN it comes to the subject of approved potable water systems, there is a great deal to learn. Unlike code requirements for health care facilities and institutional plumbing, areas in which not all plumbers get involved, the rules and regulations that pertain to potable water supplies affect all plumbers, even service and repair plumbers. As a plumber, there is no good way to avoid working with potable water supplies.

We have just completed a long chapter on vents and venting, and we are about to embark on yet another long chapter. The added length is necessary to cover the pertinent aspects of installing and working with potable water systems. If you are tired from working through the last chapter, take a break before you jump into this one. You will absorb more if you have a clear head and rested eyes.

Potable water is, of course, another name for drinking water. Good, clean water is essential to life, and it is the job of plumbers to install systems that supply potable water. If plumbers make installation mistakes or choose the wrong materials, serious health problems can arise. If you make a mistake that creates a health problem for your customer, you could be sued. Your employer could be sued. What's worse, because of your mistake someone might suffer from illness, serious burns, or even death. Potable water is not a subject to be taken lightly.

You can rest assured that numerous questions on the licensing exam will pertain to potable water. If you are not knowledgeable about water distribution systems, there will be enough questions on the exam to make it difficult for you to pass. Before you start the multiple-choice exam, review Tables 13-1 through 13-3 at the end of this chaper for sizing information. As I mentioned earlier, this is a long chapter, so let's get right to the multiple-choice exam.

Multiple-choice exam

1) Water that is not safe for drinking is called:

A) Potable water
B) Non-potable water
C) Either A or B
D) Neither A nor B

2) Potable water may be used for which of the following purposes:

A) Personal use
B) Culinary use
C) Both A and B
D) Neither A nor B

3) A valve that is pressure-actuated and held closed by a spring or other means and designed to relieve pressure automatically at a preset point is called:

A) A relief valve
B) Pressure-reducing valve
C) A vacuum breaker
D) None of the above

4) A device used to absorb pressure surges that occur when a flow of water is stopped rapidly is known as:

A) A pressure-reducing valve
B) A vacuum breaker
C) A water-hammer arrestor
D) None of the above

5) When a valve can be opened to its full-open position and provide a passageway with a diameter equal to that of the pipe in which the valve is installed, the valve is called:

A) A stop valve
B) A tempering valve
C) A full-open valve
D) None of the above

6) What type of valve combines to protect against excessive temperature and pressure build-ups?

A) A temperature-and-pressure relief valve
B) A temperature relief valve

C) A tempering valve
D) None of the above

7) A valve or faucet that closes automatically when released manually is called a:

A) Timed shut-off
B) A stop valve
C) A quick-closing valve
D) None of the above

8) A valve or faucet that is controlled by a mechanical means for fast-action closing is called a:

A) Supply valve
B) A stop valve
C) A quick-closing valve
D) None of the above

9) Potable water must be supplied for which of the following uses:

A) Drinking
B) Bathing
C) Culinary uses
D) All of the above

10) A pipe that supplies water to branches and that extends at least one full story is called:

A) A fixture supply
B) An artery
C) A riser
D) A conductor

11) The pressure of the water supply near a faucet while the faucet is wide open and flowing at full force is known as:

A) Street pressure
B) Flow pressure
C) Static pressure
D) Controlled pressure

12) A pipe that runs from a source of potable water to the water distribution system of a building is known as:

A) A water main
B) A well pipe
C) A main
D) A water service

13) Which of the following are required to make up a water supply system:

A) Water distribution pipes
B) A water service pipe
C) Both A and B
D) A check valve

14) Hot water is required in buildings where:

A) People have permanent residences
B) Livestock is maintained
C) Both A and B
D) Either A or B

15) Hot water is required in buildings where:

A) People have permanent residences
B) Livestock is maintained
C) People are employed
D) Both A and C

16) Cold water is required in buildings where:

A) Plumbing is installed
B) People are employed
C) People maintain permanent residences
D) All of the above

17) Potable water is required for which of the following:

A) Food processing
B) Medical and pharmaceutical processing
C) Culinary purposes
D) All of the above

18) The minimum diameter of a water service pipe is:

A) ½ inch
B) ¾ inch
C) 1 inch
D) 1¼ inches

19) The minimum size of a fixture water supply pipe for a bidet is:

A) ⅜ inch
B) ½ inch
C) ¾ inch
D) 1 inch

20) The minimum size of a fixture water supply pipe for a bathtub is:

A) ⅜ inch
B) ½ inch
C) ¾ inch
D) 1 inch

21) The minimum size of a fixture water supply pipe for a combination sink and tray is:

A) ⅜ inch
B) ½ inch
C) ¾ inch
D) 1 inch

22) The minimum size of a fixture water supply pipe for a domestic dishwasher is:

A) ⅜ inch
B) ½ inch
C) ¾ inch
D) 1 inch

23) The minimum size of a fixture water supply pipe for a drinking fountain is:

A) ⅜ inch
B) ½ inch
C) ¾ inch
D) 1 inch

24) The minimum size of a fixture water supply pipe for a hose bibb is:

A) ⅜ inch
B) ½ inch
C) ¾ inch
D) 1 inch

25) The minimum size of a fixture water supply pipe for a kitchen sink is:

A) ⅜ inch
B) ½ inch
C) ¾ inch
D) 1 inch

26) The minimum size of a fixture water supply pipe for a laundry tray with up to 3 compartments is:

A) ⅜ inch
B) ½ inch
C) ¾ inch
D) 1 inch

27) The minimum size of a fixture water supply pipe for a lavatory is:

A) ⅜ inch
B) ½ inch
C) ¾ inch
D) 1 inch

28) The minimum size of a fixture water supply pipe for a single shower head is:

A) ⅜ inch
B) ½ inch
C) ¾ inch
D) 1 inch

29) The minimum size of a fixture water supply pipe for a service sink is:

A) ⅜ inch
B) ½ inch
C) ¾ inch
D) 1 inch

30) The minimum size of a fixture water supply pipe for a flush-tank urinal is:

A) ⅜ inch
B) ½ inch
C) ¾ inch
D) 1 inch

31) The minimum size of a fixture water supply pipe for a flush-valve urinal is:

A) ⅜ inch
B) ½ inch
C) ¾ inch
D) 1 inch

32) The minimum size of a fixture water supply pipe for a wall hydrant is:

A) ⅜ inch
B) ½ inch
C) ¾ inch
D) 1 inch

33) The minimum size of a fixture water supply pipe for a flush-tank toilet is:

A) ⅜ inch
B) ½ inch
C) ¾ inch
D) 1 inch

34) The minimum size of a fixture water supply pipe for a flush-valve toilet is:

A) ⅜ inch
B) ½ inch
C) ¾ inch
D) 1 inch

35) A water service pipe and a building sewer must be separated by:

A) A masonry barrier
B) Undisturbed earth
C) Compacted earth
D) Either B or C

36) The bottom of a water service pipe must never be closer than _____ inches from the top of a sewer.

A) 6
B) 12
C) 18
D) 24

37) If a water service pipe is installed in a common trench with a sewer, the water service pipe must be placed to one side of the trench and installed on:

A) An even grade
B) A solid shelf of earth
C) A pitch of ¼ inch per foot
D) None of the above

38) Potable water service pipes may not be installed in, under, or above:

A) Cesspools
B) Septic tanks
C) Septic discharge fields
D) All of the above

39) Caution must be exercised and devices might be required to prevent the contents of hot and cold water pipes from:

A) Flowing between the two piping systems
B) Losing their ambient temperature
C) Both A and B
D) Neither A nor B

40) Fixture supply pipes must not terminate at distances greater than _____ inches from the point of connection with the fixture. Fill in the blank.

A) 12
B) 16
C) 30
D) 32

41) Fixture supply pipes must extend to the floor or _____ adjacent to the fixture.

A) The next fixture
B) Wall
C) Ceiling
D) Any of the above

42) An approved pressure-reducing valve must be installed in a water service pipe of a building, near where the pipe enters a building, when the street pressure of the water main providing potable water exceeds _____ pounds per square inch. Fill in the blank.

A) 60
B) 75
C) 80
D) 100

43) Pressure-reducing valves, as described in question 42, must reduce the water pressure entering the water distribution system to a point not to exceed _____ pounds per square inch. Fill in the blank.

A) 60
B) 75
C) 80
D) 100

44) Under normal conditions, a water hammer arrestor is required wherever certain valves are used. Which of the following valves are of a type to require such a device?

A) Gate valves
B) Stop-and-waste valves
C) Quick-closing valves
D) All of the above

45) Water-hammer arrestors are required to be installed so that they are:

A) Accessible
B) Readily accessible
C) Not concealed in any way
D) Easy to clean

46) When the pressure of a water main that supplies potable water to a building fluctuates, the water distribution system in the building must be designed for the _____ pressure available. Fill in the blank.

A) Minimum
B) Maximum
C) Average
D) Consolidated

47) When manifolds are used for the distribution of potable water, the cut-off valves located at the manifolds must be:

A) Full-open valves
B) Full-flow valves
C) Gate valves
D) Labeled to identify the fixture being served by the valve

48) Where manifolds are installed, they must be:

A) Accessible
B) Readily accessible
C) 42 inches above the finished floor
D) 60 inches above the finished floor

49) Water consumption for water closets that are exempt from normal water-saving rules shall not exceed _____ gallons per flushing cycle. Fill in the blank.

A) 3.5
B) 3.6
C) 4
D) 5

50) Water consumption for urinals that are exempt from normal water-saving rules shall not exceed _____ gallons per flushing cycle. Fill in the blank.

A) 1.5
B) 2.5
C) 3.5
D) 3.6

51) All full-open valves must be:

A) Gate valves
B) Accessible
C) Readily accessible
D) Labeled

52) The maximum amount of water consumption for toilets in a majority of locations shall be no more than _____ gallons per flushing cycle.

A) 1.5
B) 1.6
C) 3.5
D) 4

53) The maximum amount of water consumption for urinals in a majority of locations shall be no more than _____ gallons per flushing cycle. Fill in the blank.

A) 1
B) 1.5
C) 1.6
D) 2.5

54) When the water pressure is 80 pounds per square inch, the maximum flow rate for a single shower head shall not exceed _____ gallons per minute. Fill in the blank.

A) 1.5
B) 1.75
C) 2
D) 2.5

55) When the water pressure is 80 pounds per square inch, the maximum flow rate for a public lavatory shall not exceed _____ gallons per minute. Fill in the blank.

A) 0.5
B) 1
C) 1.25
D) 1.5

56) When the water pressure is 60 pounds per square inch, the maximum flow rate for a private lavatory shall not exceed _____ gallons per minute. Fill in the blank.

A) 1.5
B) 1.75
C) 2
D) 2.2

57) When the water pressure is 60 pounds per square inch, the maximum flow rate for a sink faucet shall not exceed _____ gallons per minute. Fill in the blank.

A) 1.5
B) 1.75
C) 2
D) 2.2

58) The maximum flow rate for a public lavatory with self-closing valves shall not exceed _____ gallons per minute. Fill in the blank.

A) 0.25
B) 0.5
C) 0.75
D) 1

59) What type of valve is required near the curb when a building water service pipe connects to a public water supply?

A) Stop-and-waste valve
B) Stop valve
C) Full-open valve
D) Compression valve

60) When a water service becomes a water distribution main, which of the following types of valves is required?

A) Stop-and-waste valve
B) Stop valve
C) Full-open valve
D) Compression valve

61) A full-open valve is required on the discharge side of every _____. Fill in the blank.

A) Fixture
B) Sink
C) Water meter
D) Water heater

62) A pipe that provides a water supply to a gravity or pressurized water tank is required to be fitted with _____. Fill in the blank.

A) A stop-and-waste valve
B) A stop valve
C) A full-open valve
D) A compression valve

63) A full-open valve is required on the supply side of every _____. Fill in the blank.

A) Fixture
B) Sink
C) Dishwasher
D) Water heater

64) Which of the following cut-off valves may be used with a sillcock?

A) Stop-and-waste valve
B) Full-open valve
C) Gate valve
D) Any of the above

65) A cut-off valve is required on a pipe that supplies water to which of the following:

A) Appliances
B) Mechanical equipment
C) Sillcocks
D) All of the above

66) When valves are installed in locations that are not adjacent to the fixture or appliance being served, the valves must be _____. Fill in the blank.

A) Full-open valves
B) Full-flow valves
C) Identified and labeled to show the fixture being served
D) Both A and C

67) Service and hose-bibb valves must be _____. Fill in the blank.

A) Full-open valves
B) Full-flow valves
C) Identified and labeled to show the fixture being served
D) Both A and C

68) Combination stop-and-waste valves are not allowed to be installed _____. Fill in the blank.

A) In finished walls
B) In ceilings
C) Underground
D) All of the above

69) All potable water openings and outlets must be protected against _____. Fill in the blank.

A) Rodent invasion
B) Insect invasion
C) Vandalism
D) Back flow

70) To be considered hot water, the temperature of the water must be at least ____ degrees Fahrenheit. Fill in the blank.

A) 100
B) 110
C) 120
D) 140

This concludes the multiple-choice test on potable water systems. We are about to begin the true-false testing on this subject. These tests will be more extensive than the ones in earlier chapters. You might want to take a short break before you continue. When you are ready, proceed to the true-false test that applies to your region.

True-false test for zone 1

1) It is a violation of the plumbing code to make a connection between a potable water supply pipe and any other pipe that is not used solely for the purpose of conveying potable water, unless suitable back-flow prevention is provided.

True False

2) It shall be a violation of the plumbing code to connect a private water supply to a public water supply, without prior approval from code officials.

True False

3) An atmospheric vacuum breaker is one that consists of a body, a checking member, and an atmospheric opening.

True False

4) A double check-valve, back-flow prevention assembly is made up of two independently acting, internally loaded check valves, two properly located test cocks, and two isolation valves.

True False

5) Fixtures with effective outlet openings of ½-inch or less, when not affected by side walls, are required to have a minimum airgap of 1 inch.

True False

6) Fixtures with effective outlet openings of ¾ inch or less, when not affected by side walls, are required to have a minimum airgap of 1¼ inch.

True False

7) Fixtures with effective outlet openings of more than 1 inch, when not affected by side walls, are required to have a minimum airgap of 2 times the diameter of the effective outlet opening.

True False

8) Over-the-rim bath fillers, when not affected by side walls, are required to have a minimum airgap of 2 inches.

True False

9) Fixtures with effective outlet openings of ½ inch or less, when affected by side walls, are required to have a minimum airgap of 1¼ inches.

True False

10) Fixtures with effective outlet openings of ¾ inch or less, when affected by side walls, are required to have a minimum airgap of 2¼ inches.

True False

11) Fixtures with effective outlet openings of more than 1 inch, when affected by side walls, are required to have a minimum airgap of 3 times the diameter of the effective outlet opening.

True False

12) Over-the-rim bath fillers, when affected by side walls, are required to have a minimum airgap of 3 inches.

True False

13) When potable water is discharged to the drainage system, it must be by means of an approved airgap of 2 pipe diameters of the supply inlets, but under no circumstances shall the gap be less than 1 inch.

True False

14) Vacuum breakers used with flushometer valves for water closets and urinals must be of an atmospheric type.

True False

15) Vacuum breakers used with flushometer valves for water closets and urinals must be installed on the discharge side of the flushometer valve.

True False

16) Back-flow preventers may not be installed in areas that contain fumes that are toxic, poisonous, or corrosive.

True False

17) Water-cooled equipment does not produce back pressure and, therefore, is not required to be equipped with back-flow protection.

True False

18) Water pipe and fittings must contain no more than 8 percent lead.

True False

19) All required shutoff valves must be readily accessible.

True False

20) When the water pressure of a potable water system is less than 25 pounds per square inch, a tank and pump must be used to boost the pressure.

True False

21) Relief valves located within a building must be trapped.

True False

22) Drains for relief valves that terminate outside a building must do so no more than 2 feet above the ground and no less than 6 inches above the ground.

True False

23) The terminal end of a drain from a relief valve must be provided with threads, in the event a hose connection is needed.

True False

24) All water service yard piping must be buried at least 6 inches below the average local frost depth.

True False

25) The minimum amount of ground cover allowed over a water service yard piping system is 12 inches.

True False

26) Unions are required at all fixture connections.

True False

27) The friction loss created by the installation of a water softener must be allowed for when sizing pipes for a water distribution system.

True False

28) When calculating the number of fixture units for a water supply outlet that is not listed in the local code book, it is assumed that a ½-inch outlet will result in a combined fixture-unit rating of 2, for private use.

True False

29) When calculating the number of fixture units for a water supply outlet that is not listed in the local code book, it is assumed that a ¾-inch outlet will result in a combined fixture-unit rating of 4, for private use.

True False

30) When calculating the number of fixture units for a water supply outlet that is not listed in the local code book, it is assumed that a 1-inch outlet will result in a combined fixture-unit rating of 6, for private use.

True False

True-false test for zone 2

1) All piping and outlets conveying non-potable water must be adequately and durably identified by a distinctive red paint so that they are readily distinguished from potable-water piping.

   True False

2) All piping and outlets conveying non-potable water must be adequately and durably identified by a distinctive yellow paint so that they are readily distinguished from potable-water piping.

   True False

3) When non-potable water is conveyed in a pipe, any valve associated with the non-potable water must be labeled with the words, "Non-potable Water."

   True False

4) All water distribution systems must be protected against back flow.

   True False

5) A vacuum breaker installed for a flushometer valve must be installed so that the critical level is not less than 6 inches above the rim of the bowl.

   True False

6) Lawn sprinkling systems are exempt from back-flow protection.

   True False

7) Fixture valve outlets with hose attachments, such as hose bibbs, must be protected by either a back-siphonage preventer or a vacuum breaker.

   True False

8) Vacuum breakers installed on hose bibbs must be of a type that cannot be removed.

   True False

9) When back-flow preventers are buried, they must be buried so that there will be no less than 12 inches of cover above them.

True False

10) Pipe-applied vacuum breakers must be installed no less than 6 inches above the flood-level rim of the fixture receptor or device being served.

True False

11) Approved deck-mounted vacuum breakers may be installed with the critical level being no less than 1 inch above the flood-level rim.

True False

12) Pressure-type back-flow preventers must be provided with stop-and-waste valves on both the inlet and the outlet side of the device.

True False

13) Pressure-type back-flow preventers must be provided with gate valves on both the inlet and the outlet side of the device.

True False

14) The minimum working pressure of plastic piping installed outside of a foundation shall be 160 pounds per square inch, at 73 degrees Fahrenheit.

True False

15) Flux used with potable water must be leadfree.

True False

16) The minimum size of a water service pipe shall not be less than 1 inch.

True False

17) Private water supplies may be connected to public water supplies so long as appropriate valves are installed.

True False

18) Drains from relief valves may not connect directly to a drainage system.

True False

19) Vacuum breakers installed in conjunction with pressure tanks used for supplying potable water must be installed on the top of the tank.

True False

20) One method of disinfecting a potable water supply involves filling the system with a solution containing 50 parts per million of available chlorine and allowing the solution to stand in the system for 6 hours before flushing the system or returning it to use.

True False

21) Water supply tanks must be provided with valved drain lines at their lowest point and discharged as an indirect waste.

True False

22) One method of disinfecting a potable water supply involves filling the system with a solution containing 100 parts per million of available chlorine and allowing the solution to stand in the system for 2 hours before flushing the system or returning it to use.

True False

23) A shutoff valve must be installed on the inlet pipe of any water heater.

True False

24) All hot-water supplies shall be installed on the right side of a fixture, unless otherwise specified by the manufacturer of the fixture.

True False

25) The minimum depth of a safety pan used in conjunction with an electric water heater shall be not less than 2 inches.

True False

26) The minimum drain size for a pipe draining a safety pan in which a water heater sits shall be not less than ¾ inch.
True False

27) The temperature of mixed water to gang showers must be controlled to prevent scalding.
True False

28) Trap primers must contain integral airgaps.
True False

29) Trap primers must be readily accessible.
True False

30) All storage tanks must be equipped with adequate drain cocks.
True False

True-false test for zone 3

1) A riser is a water supply that extends at least one full story to convey water to branches or to a group of fixtures.
True False

2) A temperature relief valve is one that protects against excessive temperature and pressure.
True False

3) All covers installed on water supply tanks must be capable of keeping out unauthorized persons, dirt, and vermin.
True False

4) A low-pressure cutoff is required for all booster pumps in a water-pressure booster system.
True False

5) Potable water inlets to gravity tanks shall be controlled by full-open valves.

True False

6) The inlet for a gravity-type water supply must terminate with an airgap of not less than 4 inches above the overflow.

True False

7) Supply tanks for potable water may not be located beneath any soil or waste pipe.

True False

8) Stop-and-waste valves may not be installed underground.

True False

9) Water utilized for equipment cooling can be returned to the potable water supply only after passing through a carbon filter.

True False

10) The water supply connection to a carbonated beverage dispenser must be protected with the use of a double check valve with an intermediate atmospheric vent.

True False

11) All back-flow preventers must be accessible.

True False

12) Bi-weekly inspections must be made on all back-flow preventers to determine if they are working properly.

True False

13) Fixtures with an effective opening of ½ inch or less, when not affected by the proximity of walls, must have a minimum airgap of 1 inch.

True False

14) Fixtures with an effective opening of ¾-inch or less, when not affected by the proximity of walls, must have a minimum airgap of 1½ inches.

True False

15) Fixtures with an effective opening of 1 inch or less, when not affected by the proximity of walls, must have a minimum airgap of 3 inches.

True False

16) Fixtures with an effective opening of ½ inch or less, when affected by the proximity of walls, must have a minimum airgap of 1¼ inches.

True False

17) Fixtures with an effective opening of ¾ inch or less, when not affected by the proximity of walls, must have a minimum airgap of 2½ inches.

True False

18) Fixtures with an effective opening of 1 inch or less, when not affected by the proximity of walls, must have a minimum airgap of 3 inches.

True False

19) A potable water connection to a boiler requires the installation of a back-flow preventer with an intermediate atmospheric vent.

True False

20) Tempered water is water with a temperature that ranges from 85 degrees to 110 degrees Fahrenheit.

True False

21) When the developed length of hot water piping exceeds 100 feet, the hot water system must be equipped to provide hot water with a temperature that could be expected from piping that was less than 100 feet in length.

True False

22) A circulating pump installed on a return circulation hot water system must be arranged to allow it to be shut off automatically or manually when the hot water system is not in operation.

True False

23) The temperature of water from a tankless water heater must be tempered to not exceed 120 degrees Fahrenheit.

True False

24) Water heaters used at pressures above atmospheric shall be rated for a minimum working pressure of 150 pounds per square inch.

True False

25) All water heaters must be protected with a temperature and pressure-relief valve.

True False

26) Piping material used as a drain for a relief valve from a water heater must be rated as being suitable to handle water temperatures of at least 210 degrees Fahrenhert.

True False

27) The terminal end of a drain from a relief valve must not terminate more than 4 inches above the floor.

True False

28) When an old water heater is replaced with a new one, the existing relief valve used with the old heater may be used with the new heater.

True False

29) Temperature relief valves used with water heaters shall be set for a maximum temperature of 175 degrees Fahrenheit.

True False

30) Valves must not be installed between a safety device and the water heater being served by the devices.

True False

This brings us to the end of the tests for potable water supplies. Compare your answers with the correct answers at the end of the chapter. Review any areas that gave you trouble, and then move on to the next chapter, where we will study private water systems.

Answers to multiple-choice exam

1) B	19) A	37) B	54) D
2) C	20) B	38) D	55) A
3) A	21) B	39) A	56) D
4) C	22) B	40) C	57) D
5) C	23) A	41) B	58) A
6) A	24) B	42) C	59) C
7) C	25) B	43) C	60) C
8) C	26) B	44) C	61) C
9) D	27) A	45) A	62) C
10) C	28) B	46) A	63) D
11) B	29) B	47) D	64) D
12) D	30) B	48) A	65) D
13) C	31) C	49) C	66) C
14) A	32) B	50) A	67) C
15) D	33) A	51) B	68) C
16) D	34) D	52) B	69) D
17) D	35) D	53) A	70) B
18) B	36) B		

Answers to true-false test for zone 1

1) T	11) T	21) F
2) T	12) T	22) T
3) T	13) T	23) F
4) F	14) T	24) F
5) T	15) T	25) T
6) F	16) T	26) F
7) T	17) F	27) T
8) T	18) T	28) T
9) F	19) F	29) F
10) T	20) F	30) T

Answers to true-false test for zone 2

1) F	11) T	21) T
2) T	12) F	22) T
3) T	13) T	23) T
4) T	14) T	24) F
5) F	15) T	25) F
6) F	16) F	26) F
7) T	17) F	27) T
8) T	18) T	28) T
9) F	19) T	29) F
10) T	20) T	30) T

Answers to true-false test for zone 3

1) T	11) T	21) T
2) F	12) F	22) T
3) T	13) T	23) F
4) T	14) T	24) F
5) F	15) F	25) T
6) T	16) F	26) T
7) T	17) T	27) T
8) T	18) T	28) F
9) F	19) T	29) F
10) T	20) T	30) T

Table 13-1 **Common Minimum Fixture-Supply Sizes**

Fixture	Minimum supply size
Bathtub	½"
Bidet	⅜"
Shower	½"

Toilet	⅜"
Lavatory	⅜"
Kitchen sink	½"
Dishwasher	½"
Laundry tub	½"
Hose bibb	½"

Sample Pressure and Pipe Chart Table 13-2

Size of water meter and street service	Size of water service and distribution pipes	Maximum length of water pipe 40'	60'	80'	100'	150'	200'
¾"	½"	9	8	7	6	5	4
¾"	¾"	27	23	19	17	14	11
¾"	1"	44	40	36	33	28	23
1"	1"	60	47	41	36	30	25
1"	1¼"	102	87	76	67	52	44

Equivalent Fixture-Unit Ratings for Sizing Table 13-3

Fixture	Fixture-unit rating combining hot and cold water
Toilet	3
Lavatory	1
Tub/shower combination	2
Shower	2
Dishwasher	2
Kitchen sink	2
Laundry hook-up	2
Sillcock	3

CHAPTER 14

Learning about private water supply systems

PLUMBERS who work in large cities might go through their entire career without needing to learn about private water supply systems. Perhaps the only time this subject will be of importance to them is on the day of the licensing exam. Other plumbers, however, often work with private water supplies. They must be knowledgeable of the code requirements for private water systems during their routine work, as well as for the licensing exam.

If you live in an area where private water supplies are rare, you might decide to skip this chapter. I hope you won't. Even though you might never work with a private system, you will probably be asked some questions about such systems on the licensing exam.

What is a private water supply? Private water supplies are sources of potable water that provide water for drinking and other plumbing needs. The sources might be wells, cisterns, or other reserves of suitable water. In general, for the purposes of this chapter, private and individual water supplies are those that do not originate from a public water main.

Before we get on with the testing, I want to clear up a bit of confusion that often arises about private or individual water supplies. An individual water supply is not, by definition, one that serves only one person, family, dwelling, or building. It is a water supply that serves one or more families that is not an approved public water supply. In other words, a well that provides water to a small group of homes is still considered an individual water supply. This point has, in the past, confused many of my students, so don't let it have a detrimental effect on your scoring during the following mock tests. Now, let's begin our testing with some multiple-choice questions.

Multiple-choice exam

1) A small, covered tank used for storing water, which is not normally used as potable water, for a home or farm is called a:

A) Well
B) Cesspool
C) Cistern
D) None of the above

2) When a pipe is fitted with a point and a screen and then driven into the ground to a point where water is found, the type of well that has been created is a:

A) Bored well
B) Drilled well
C) Driven well
D) Dug well

3) Wells that have shafts with large diameters and a casing are normally considered to be:

A) Bored wells
B) Drilled wells
C) Driven wells
D) Dug wells

4) When an auger is used to make a well and a casing is installed, the type of well created is a:

A) Bored well
B) Drilled well
C) Driven well
D) Dug well

5) When a drilling machine, of any type, is used to make a well and a casing is installed, the type of well created is a:

A) Bored well
B) Drilled well
C) Driven well
D) Dug well

6) Which of the following can be used as a private water source:

A) Bored well
B) Springs
C) Streams
D) All of the above

7) Some types of water sources might not be used as individual water supplies unless they are treated properly to prevent contamination. Which of the following types of individual water supplies require such treatment:

A) Drilled wells
B) Land cisterns
C) Streams
D) Both B and C

8) After construction or major repair, a private water supply system must be _____. Fill in the blank.

A) Purged of deleterious matter
B) Disinfected
C) Sealed
D) Both A and B

9) What type of water must be treated prior to use as an individual water supply?

A) Running water
B) Still water
C) Surface water
D) All of the above

10) There are certain distance requirements that come into effect with the installation of private water supply systems. Potable ground water sources and suction lines from pumps must not be installed closer than specific distances, as established by the plumbing code, to possible sources of contamination. How far, at a minimum, must a potable ground water source be from a sewer?

A) 5 feet
B) 10 feet
C) 100 feet
D) None of the above

11) How far, at a minimum, must a potable ground water source be from a septic tank?

A) 15 feet
B) 25 feet
C) 100 feet
D) None of the above

12) How far, at a minimum, must a potable ground water source be from a subsurface pit?

A) 15 feet
B) 25 feet
C) 100 feet
D) None of the above

13) How far, at a minimum, must a potable ground water source be from a pasture?

A) 15 feet
B) 25 feet
C) 100 feet
D) None of the above

14) How far, at a minimum, must a potable ground water source be from a subsurface disposal field?

A) 15 feet
B) 25 feet
C) 50 feet
D) 100 feet

15) How far, at a minimum, must a potable ground water source be from a seepage pit?

A) 15 feet
B) 25 feet
C) 50 feet
D) 100 feet

16) How far, at a minimum, must a potable ground water source be from a barnyard?

A) 15 feet
B) 25 feet
C) 50 feet
D) 100 feet

17) How far, at a minimum, must a potable ground water source be from a farm silo?

A) 15 feet
B) 25 feet
C) 50 feet
D) 100 feet

18) Private potable well supplies are not allowed to be developed from a water table that is less than _____ feet below the ground surface, unless the water is treated to prevent contamination. Fill in the blank.

A) 5
B) 10
C) 15
D) 25

19) Well sites must be positively drained and they must be at elevations that are _____ potential sources of contamination.

A) Higher than
B) Lower than
C) Equal to
D) Any of the above

20) The casings used to line any well for potable water must be:

A) Made of concrete
B) Made of steel
C) Watertight
D) Both B and C

21) A casing for a well must extend below the ground surface for a minimum distance of:

A) 5 feet
B) 10 feet
C) 15 feet
D) 25 feet

22) Well casing must extend a minimum of _____ inches above the well platform. Fill in the blank.

A) 4
B) 6
C) 12
D) 18

23) The casings used to line drilled or driven wells must be:

A) Made of concrete
B) Made of steel
C) Made of stainless steel
D) None of the above

24) The annular space between the earth and the outside of a casing installed for a drilled or driven well must be filled with cement grout to a minimum distance of _____ feet below the ground surface. Fill in the blank.

A) 5
B) 10
C) 15
D) 25

25) Which of the following materials is acceptable for use as a casing for a dug or bored well?

A) Concrete
B) Tile
C) Galvanized or corrugated metal pipe
D) All of the above

26) Well casings for dug or bored wells must be surrounded by _____ inches of grout poured into the hole between the outside of the casing and the earth.

A) 3
B) 4
C) 6
D) 8

27) The casing for a dug or bored well must extend at least ____ feet below the ground surface. Fill in the blank.

A) 5
B) 10
C) 15
D) 25

28) All covers for potable water wells must be:

A) Made of steel
B) Made of concrete
C) Watertight
D) Permanently sealed

29) Well covers for potable water wells must extend at least _____ inches over the outside of the well casing or wall. Fill in the blank.

A) 2
B) 4
C) 6
D) 12

30) All potable water wells and springs shall be built in such a way so that surface drainage is:

A) Absorbed
B) Retained
C) Diverted
D) Unaffected

31) When a well is located in a flood-hazard zone or a high-hazard zone, the top of the well casing or pipe sleeve must be extended at least _____ inches above the base flood elevation, or shall have a sealed cover capable of resisting hydrostatic and hydrodynamic loads and stresses, including the effects of buoyancy, during a flood. Fill in the blank.

A) 4
B) 8
C) 12
D) 18

32) When water pumps are installed in basements, the pumps must be installed on blocks or shelves that will place the pump at least _____ inches above the basement floor. Fill in the blank.

A) 4
B) 8
C) 12
D) 18

33) Water pumps must be designed to maintain a _____. Fill in the blank.

A) Constant water temperature
B) Prime
C) Both A and B
D) Neither A nor B

34) Dug wells must be equipped with covers that are fitted with sleeves that allow the pump, pump-suction pipe, cylinder, or jet body to be removed without _____. Fill in the blank.

A) Disturbing the well cover
B) The use of a winch
C) The use of a pump puller
D) None of the above

35) Any penetrations in well casings must be sealed and made:

A) Watertight
B) With hole saws
C) Both A and B
D) Neither A nor B

I promised you a short chapter, and you got one. This is the end of the test on private water supplies. There are no true-false tests in this chapter. Why is the chapter so brief? The plumbing codes in zones 1 and 2 do not address the issue of private water supplies with the same aggressiveness that the code for zone 3 does. For this reason, plumbers in zones 1 and 2 will not find many of the questions or answers dealt with in this chapter in their local code books.

Since your mind is fresh from such a short chapter, check your answers against the correct answers at the end of this chapter, and then move on to chapter 15, where we will talk about pipe hangers and supports. This is an issue all major codes dictate and enforce.

Answers to multiple-choice exam

1) C
2) C
3) D
4) A
5) B
6) D
7) D
8) D
9) C
10) B
11) B
12) D
13) C
14) C
15) C
16) D
17) B
18) B
19) A
20) C
21) B
22) B
23) B
24) B
25) D
26) C
27) B
28) C
29) A
30) C
31) C
32) D
33) B
34) A
35) A

CHAPTER 15

The principles of pipe hangers and supports

THE principles of pipe hangers and supports apply to all plumbers. You've just finished a chapter that dealt with private water supplies. As you saw in that chapter, not all plumbing codes deal equally with all issues. However, you will find that all three of the codes are concerned with the installation of pipe hangers and supports.

Plumber's helpers and apprentices are often assigned the task of installing pipe hangers and digging ditches. Both of these jobs are regulated by the code under the section on hangers and supports. If a ditch isn't prepared properly for a pipe installation, the job can fail the inspection by the code officer. If hangers are installed too far apart, a rejected inspection certificate is likely. Everyone involved with plumbing, from helpers to masters, must be aware of the code requirements for hangers and support.

Some plumbers feel that the code requirements for pipe hangers are too rigid. They think the use of hangers every four feet on plastic pipe is a waste of hangers. But the code has set the requirements with good intention. Hangers do more than hold pipes up off the ground or floor. They keep sags out of drain pipes that might result in a clogged drain. The hangers support water pipes so that they don't bang and rattle so much, and they remove stress from solder joints that might be ruptured without the prescribed hangers. Refer to Tables 15-1 through 15-14 at the end of this chapter for pipe-support intervals per zone.

Have you ever looked at a plumbing installation and seen a copper water pipe held in place with a bent-over nail? This type of hanger is prohibited by the plumbing code, and the reason for the rule is a good one. A nail can damage copper tubing in two ways. A chemical reaction can occur between the nail and the copper that will result in a leak. If the water pipe vibrates when water runs through it, the rubbing action created by the vibration can allow the nail to slowly rub a hole in the pipe.

I know a lot of plumbers resent the plumbing code, but the code exists to help and protect people and their plumbing. You will do better in your career as a professional plumber if you take the time to learn the code and apply it in your daily work. Now, let's get on with the multiple-choice exam pertaining to hangers and supports.

Multiple-choice exam

1) When earthquake loads are applicable to the hangers and supports used with plumbing pipes, _____ forces must be taken into consideration. Fill in the blank.

A) Wind
B) Water
C) Seismic
D) None of the above

2) Hangers and strapping materials used to hang and support plumbing pipes must not allow _____ action to occur. Fill in the blank.

A) Galvanic
B) Tyro
C) Swaying
D) None of the above

3) Which of the following materials may be used to support the bases of cast-iron stacks:

A) Concrete
B) Bricks laid in sand
C) Both A and B
D) Neither A nor B

4) Which of the following materials may be used to support the bases of cast-iron stacks:

A) Concrete
B) Bricks laid in mortar
C) Both A and B
D) Neither A nor B

5) Which of the following materials may be used to support the bases of cast-iron stacks:

A) Metal brackets attached to the building construction
B) Plastic brackets attached to the building construction
C) Both A and B
D) Neither A nor B

6) Anchorage shall be provided to restrain drainage piping joined with friction couplings from which of the following types of movement:

A) Aquatic
B) Lang
C) Axial
D) All of the above

7) Rigid support sway bracing shall be used at changes in directions of greater than 45 degrees for pipes with diameter of _____ inches or larger. Fill in the blank.

A) 2
B) 3
C) 4
D) 6

8) Restraints must be provided for drain pipes connected with _____ joints. Fill in the blank.

A) Lead
B) Soldered
C) Mechanical
D) All of the above

9) Restraints must be provided for drain pipes connected with _____ joints. Fill in the blank.

A) Lead
B) Screwed
C) Mechanical
D) All of the above

10) The maximum horizontal spacing of hangers installed on ABS pipe is:

A) 4 feet
B) 6 feet
C) 10 feet
D) 12 feet

11) The maximum horizontal spacing of hangers installed on aluminum tubing is:

A) 4 feet
B) 6 feet
C) 10 feet
D) 12 feet

12) The maximum horizontal spacing of hangers installed on brass pipe is:

A) 4 feet
B) 6 feet
C) 10 feet
D) 12 feet

13) The maximum horizontal spacing of hangers installed on cast-iron pipe where the pipe length is 10 feet is:

A) 4 feet
B) 6 feet
C) 10 feet
D) 12 feet

14) The maximum horizontal spacing of hangers installed on copper tubing with a diameter of 1¼ inches or less is:

A) 4 feet
B) 6 feet
C) 10 feet
D) 12 feet

15) The maximum horizontal spacing of hangers installed on copper tubing with a diameter of 1½ inches or more is:

A) 4 feet
B) 6 feet
C) 10 feet
D) 12 feet

16) The maximum horizontal spacing of hangers installed on CPVC pipe with a diameter of 1 inch or less is:

A) 2 feet
B) 3 feet
C) 4 feet
D) 6 feet

17) The maximum horizontal spacing of hangers installed on CPVC pipe with a diameter of 1¼ inches or more is:

A) 2 feet
B) 3 feet
C) 4 feet
D) 6 feet

18) The maximum horizontal spacing of hangers installed on galvanized steel pipe is:

A) 4 feet
B) 6 feet
C) 10 feet
D) 12 feet

19) The maximum horizontal spacing of hangers installed on PVC pipe is:

A) 4 feet
B) 6 feet
C) 10 feet
D) 12 feet

20) The maximum horizontal spacing of hangers installed on PB pipe is:

A) Continuous
B) 16 inches
C) 32 inches
D) 48 inches

21) The maximum horizontal spacing of hangers installed on lead pipe is:

A) Continuous
B) 16 inches
C) 32 inches
D) 48 inches

22) The maximum vertical spacing of hangers installed on PB pipe is:

A) 4 feet
B) 6 feet
C) 8 feet
D) 10 feet

23) The maximum vertical spacing of hangers installed on PVC pipe is:

A) 4 feet
B) 6 feet
C) 8 feet
D) 10 feet

24) The maximum vertical spacing of hangers installed on ABS pipe is:

A) 4 feet
B) 6 feet
C) 8 feet
D) 10 feet

25) The maximum vertical spacing of hangers installed on aluminum tubing is:

A) 4 feet
B) 10 feet
C) 12 feet
D) 15 feet

26) The maximum vertical spacing of hangers installed on brass pipe is:

A) 4 feet
B) 10 feet
C) 12 feet
D) 15 feet

27) The maximum vertical spacing of hangers installed on cast-iron pipe is:

A) 4 feet
B) 10 feet
C) 12 feet
D) 15 feet

28) The maximum vertical spacing of hangers installed on galvanized steel pipe is:

A) 4 feet
B) 10 feet
C) 12 feet
D) 15 feet

29) The maximum vertical spacing of hangers installed on copper tubing with a diameter of 1¼ inches or less is:

A) 4 feet
B) 10 feet
C) 12 feet
D) 15 feet

30) The maximum vertical spacing of hangers installed on copper tubing with a diameter of 1½ inches or more is:

A) 4 feet
B) 10 feet
C) 12 feet
D) 15 feet

31) The maximum horizontal spacing of hangers installed on cast-iron pipe where the pipe length is 5 feet is:

A) 4 feet
B) 5 feet
C) 10 feet
D) 12 feet

32) The maximum vertical spacing of hangers installed on CPVC pipe with a diameter of 1 inch or less is:

A) 5 feet
B) 10 feet
C) 12 feet
D) 15 feet

33) The maximum vertical spacing of hangers installed on CPVC pipe with a diameter of 1¼ inches or more is:

A) 5 feet
B) 6 feet
C) 8 feet
D) 10 feet

34) Concrete can be used as a base support for which of the following types of pipe:

A) Cast-iron
B) Copper
C) Either A or B
D) Neither A nor B

35) When running water distribution pipes in bundles, from a manifold, the hot and cold water _____ be grouped together in the same bundle.

A) May
B) May not
C) Must
D) None of the above

True-false test for zone 1

1) When threaded pipe, with a diameter of ¾ inch or less, is run vertically, it must be supported at intervals not to exceed once at every story.

   True False

2) When threaded pipe, with a diameter of ¾ inch or less, is run horizontally, it must be supported at intervals not to exceed 12 feet.

   True False

3) When threaded pipe, with a diameter of 1 inch or more, is run vertically, it must be supported at intervals not to exceed once at every story.

   True False

4) When threaded pipe, with a diameter of 1 inch or more, is run horizontally, it must be supported at intervals not to exceed 12 feet.

   True False

5) Copper tubing, with a diameter of 1½ inches or less, must be supported at every story, not to exceed 10 feet, when run vertically.

   True False

6) Copper tubing, with a diameter of 1½ inches or less, must be supported at intervals not to exceed 10 feet when run horizontally.

   True False

7) Copper tubing, with a diameter of 2 inches or more, must be supported at every story, not to exceed 10 feet, when run vertically.

   True False

8) Copper tubing, with a diameter of 2 inches or more, must be supported at intervals not to exceed 10 feet when run horizontally.

   True False

9) Copper tubing may be supported with nails.

True False

10) Copper tubing may not be supported with corrosive hangers.

True False

True-false test for zone 2

1) Vertical piping must be secured often enough to maintain the proper alignment of the pipe and carry the weight of the pipe and contents.

True False

2) Cast-iron stacks must be supported at their bases.

True False

3) Cast-iron pipes that are installed vertically must be supported at each story level, at intervals that do not exceed 10 feet.

True False

4) Threaded pipe must be supported at its base when run vertically.

True False

5) Threaded pipe that is run vertically must be supported at each story level, at intervals that do not exceed 20 feet.

True False

6) Copper tubing with a diameter of 1½ inches or more must be supported at each story level when installed vertically.

True False

7) Copper tubing with a diameter of 1¼ inches or less, when run vertically, must be supported at intervals that do not exceed 6 feet.

True False

8) Lead pipe must be supported at intervals that do not exceed 4 feet when the piping is installed vertically.

True False

9) Plastic piping that is installed vertically must be supported at each story for pipes with diameters of 2 inches or more.

True False

10) Plastic piping that is installed vertically must be supported at each story for pipes with diameters of 1½ inches or less.

True False

11) Borosilicate glass pipe with diameters of 3 inches, 4 inches, and 6 inches must be supported at every floor when installed vertically.

True False

12) Borosilicate glass pipe with a diameter of 2 inches or less must be supported at every other floor level when installed vertically.

True False

13) Supports for borosilicate glass pipe must be padded riser clamps with ¼-inch-thick elastomer padding that restricts sideward and downward movement.

True False

14) Aluminum pipe must be supported at its base when installed vertically.

True False

15) Aluminum pipe must be supported at each story level, at intervals not to exceed 12 feet when installed vertically.

True False

16) Cast-iron pipe, when installed horizontally, must be supported at 5-foot intervals when the sections of pipe are 5 feet long.

True False

17) Cast-iron pipe, when installed horizontally, must be supported at 5-foot intervals when the sections of pipe are 10 feet long.

True False

18) Hangers applied to cast-iron pipes, when run horizontally, must be as close to the hubs as possible.

True False

19) Threaded pipe that is installed horizontally must be supported at intervals not to exceed 10 feet.

True False

20) Copper tubing with a diameter of 1½ inches or less, when run horizontally, must be supported at intervals not to exceed 4 feet.

True False

21) Copper tubing with a diameter of 2 inches or more, when run horizontally, must be supported at intervals not to exceed 10 feet.

True False

22) Lead pipe, installed horizontally, must be supported for its entire length.

True False

23) Aluminum pipe that is installed horizontally must be supported at intervals not to exceed 12 feet.

True False

24) Borosilicate glass pipe, run horizontally, can be supported at intervals of between 8 and 10 feet.

True False

25) Unless there are more than two joints in an 8- to 10-foot section of horizontal borosilicate glass pipe, supports may not be installed closer than 8 feet apart.

True False

26) The bases of all soil, waste, and vent stacks must be supported.

True False

27) Vertical piping must be anchored in a way so that the weight load is relieved from the stack.

True False

28) Piping in masonry walls must be installed in chases or recesses which permit access to the piping for repairs and replacement.

True False

29) Aluminum pipe must be hung with padded clevis hangers when run horizontally.

True False

30) Lead pipe may be supported by strips when it is installed horizontally.

True False

True-false test for zone 3

1) Hangers, straps, and supports must support only the weight of the pipes being installed.

True False

2) Cast-iron stacks must be supported at their bases.

True False

3) Drainage piping that is joined with friction couplings must be restrained with anchorage.

True False

4) PVC pipe must be secured or supported at 4-foot intervals whether it is installed horizontally or vertically.

True False

5) Supports are often required when pipes make a change in direction.

True False

6) Pipes joined with hubless couplings must be restrained from axial movement.

True False

7) 10-foot sections of cast-iron pipe must be supported at 5-foot intervals when installed horizontally.

True False

8) Piping bundles for manifold systems must adhere to the same support and hanger regulations as provided for single pipes of the same materials.

True False

9) Metal brackets are an approved support for the bases of cast-iron stacks.

True False

10) CPVC pipe, with a diameter of 1 inch or less, requires intervals of support that do not exceed 4 feet when the piping is installed horizontally.

True False

You have just completed all of the tests for hangers and supports. This chapter has not been long, but the contents are important. Failure to comply with the rules and regulations that pertain to hangers and supports can result in rejected inspections, broken pipes, noisy pipes, and dissatisfied customers.

If you fail to learn the rules that apply to pipe supports, your score on the licensing exam will suffer. Take some time to review your work. Check your answers with those at the end of this chapter. When you find a question that you answered incorrectly, consult your code book to learn why your answer was wrong. Don't feel bad if you didn't do well on these tests. Much of this work is memory work, and if you have not prepared with the help of your code book, you should not expect to get a great score on these mock tests. The important thing is to spend enough time now to learn the lessons, so that you will do well on the licensing exam.

We are now ready to move onto the next chapter. Chapter 16 is going to test your knowledge of plumbing fixtures. You might be surprised at how complicated it can be to pick the proper fixtures for a job. When you're ready, turn to the next chapter and continue your exam preparation.

Answers to multiple-choice exam

1) C	10) A	19) A	28) D
2) A	11) C	20) C	29) B
3) A	12) C	21) A	30) B
4) C	13) C	22) A	31) B
5) C	14) B	23) A	32) A
6) C	15) C	24) A	33) B
7) C	16) B	25) D	34) A
8) C	17) C	26) B	35) B
9) C	18) D	27) D	

Answers to true-false test for zone 1

1) T	5) T	8) T
2) F	6) F	9) F
3) T	7) T	10) T
4) T		

Answers to true-false test for zone 2

1) T	6) T	11) T
2) T	7) F	12) T
3) F	8) T	13) T
4) T	9) T	14) T
5) F	10) F	15) F

16) T	21) T	26) T
17) F	22) T	27) T
18) T	23) F	28) T
19) F	24) T	29) F
20) F	25) T	30) A

Answers to true-false test for zone 3

1) F	5) T	8) T
2) T	6) T	9) T
3) T	7) F	10) F
4) T		

Table 15-1 **Horizontal Pipe-Support Intervals in Zone 1**

Type of drainage pipe	Maximum distance of supports
ABS	4'
Cast iron	At each pipe joint*
Galvanized (1" and larger)	12'
Galvanized (¾" and smaller)	10'
PVC	4'
Copper (2" and larger)	10'
Copper (1½" and smaller)	6'

*Note—cast-iron pipe must be supported at each joint, but supports may not be more than 10 feet apart.

Table 15-2 **Horizontal Pipe-Support Intervals in Zone 1**

Type of vent pipe	Maximum distance of supports
ABS	4'
Cast iron	At each pipe joint*
Galvanized	12'

Copper (1½" and smaller)	6'
PVC	4'
Copper (2" and larger)	10'

*Note—cast-iron pipe must be supported at each joint, but supports may not be more than 10 feet apart.

Table 15-3

Vertical Pipe-Support Intervals in Zone 1**

Type of drainage pipe	Maximum distance of supports
Lead pipe	4'
Cast iron	At each story
Galvanized	At least every other story
Copper	At each story*
PVC	Not mentioned
ABS	Not mentioned

*Note—support intervals may not exceed 10 feet.

**Note—all stacks must be supported at their bases.

Table 15-4

Vertical Pipe-Support Intervals in Zone 1**

Type of vent pipe	Maximum distance of supports
Lead pipe	4'
Cast iron	At each story
Galvanized	At least every other story
Copper	At each story*
PVC	Not mentioned
ABS	Not mentioned

*Note—support intervals may not exceed 10 feet.

**Note—all stacks must be supported at their bases.

Table 15-5 **Support Intervals for Supporting Water Pipe in Zone 1**

Type of pipe	Vertical support interval	Horizontal support interval
Threaded pipe (¾" and smaller)	Every other story	10'
Threaded pipe (1" and larger)	Every other story	12'
Copper tube (1½" and smaller)	Every story, not to exceed 10'	6'
Copper tube (2" and larger)	Every story, not to exceed 10'	10'
Plastic pipe	Not mentioned	4'

Table 15-6 **Horizontal Pipe-Support Intervals in Zone 2**

Type of drainage pipe	Maximum distance of supports
ABS	4'
Cast iron	At each pipe joint
Galvanized (1" and larger)	12'
PVC	4'
Copper (2" and larger)	10'
Copper (1½" and smaller)	6'

Table 15-7 **Horizontal Pipe-Support Intervals in Zone 2**

Type of vent pipe	Maximum distance of supports
ABS	4'
Cast iron	At each pipe joint
Galvanized	12'
PVC	4'
Copper (2" and larger)	10'
Copper (1½" and smaller)	6'

Table 15-8

Vertical Pipe-Support Intervals in Zone 2***

Type of vent pipe	Maximum distance of supports
Lead pipe	4'
Cast iron	At each story*
Galvanized	At each story**
Copper (1¼")	4'
Copper (1½" and larger)	At each story
PVC (1½" and smaller)	4'
PVC (2" and larger)	At each story
ABS (1½" and smaller)	4'
ABS (2" and larger)	At each story

*Note—support intervals may not exceed 15 feet.

**Note—support intervals may not exceed 30 feet.

***Note—all stacks must be supported at their bases.

Table 15-9

Vertical Pipe-Support Intervals in Zone 2***

Type of drainage pipe	Maximum distance of supports
Lead pipe	4'
Cast iron	At each story*
Galvanized	At each story**
Copper (1¼" and smaller)	4'
Copper (1½" and larger)	At each story
PVC (1½" and smaller)	4'
PVC (2" and larger)	At each story
ABS (1½" and smaller)	4'
ABS (2" and larger)	At each story

*Note—support intervals may not exceed 15 feet.

**Note—support intervals may not exceed 30 feet.

***Note—all stacks must be supported at their bases.

Table 15-10 **Support Intervals for Supporting Water Pipe in Zone 2**

Type of pipe	Vertical support interval	Horizontal support interval
Threaded pipe	30'	12'
Copper tube (1¼" and smaller)	4'	6'
Copper tube (1½")	Every story	6'
Copper tube (larger than 1½")	Every story	10'
Plastic pipe (2" and larger)	Every story	4'
Plastic pipe (1½" and smaller)	4'	4'

Table 15-11 **Horizontal Pipe-Support Intervals in Zone 3**

Type of vent pipe	Maximum distance of supports
Lead pipe	Continuous
Cast iron	5'
Galvanized	12'
Copper tube (1¼")	6'
Copper tube (1½" and larger)	10'
ABS	4'
PVC	4'
Brass	10'
Aluminum	10'

Table 15-12 **Horizontal Pipe-Support Intervals in Zone 3**

Type of vent pipe	Maximum distance of supports
PB pipe	32"
Lead pipe	Continuous
Cast iron	5' or at each joint

Galvanized	12'
Copper tube (1¼")	6'
Copper tube (1½" and larger)	10'
ABS	4'
PVC	4'
Brass	10'
Aluminum	10'

Vertical Pipe-Support Intervals in Zone 3 Table 15-13

Type of vent pipe	Maximum distance of supports
Lead pipe	4'
Cast iron	15'
Galvanized	15'
Copper tubing	10'
ABS	4'
PVC	4'
Brass	10'
PB pipe	4'
Aluminum	15'

Vertical Pipe-Support Intervals in Zone 3 Table 15-14

Type of vent pipe	Maximum distance of supports
Lead pipe	4'
Cast iron	15'
Galvanized	15'
Copper tubing	10'
ABS	4'
PVC	4'
Brass	10'
Aluminum	15'

CHAPTER 16

Choosing and installing plumbing fixtures

THE job of choosing and installing proper plumbing fixtures is more complicated than many people realize. A lot of plumbers think that fixture choice is up to the customer. To some extent, this is true, but you must be able to evaluate the customer's request to make sure that the fixture is appropriate to the system and will work in the location in which it is to be installed.

Master plumbers and architects are typically responsible for determining the minimum fixture requirements for a building. If you are thinking in terms of residential plumbing, the required fixtures are not difficult to list. However, when you get involved in commercial plumbing, knowing what types of fixtures are required and in what numbers they must be installed gets more difficult. To expand on this, let me give you a few examples of what a master plumber must be able to determine.

As an example, let's assume that you are asked to design a plumbing system for a factory. Would you know that you must have one water closet for each 100 people in the factory? Did you know that one lavatory would be required for every 40 people in the building? There are no bathtubs required for a factory, but there must be at least one drinking fountain for every 100 people in the factory. There is also a requirement that dictates that one service sink be installed. Now, let's compare this fixture selection to that of a building used as a dormitory.

Dormitories must be equipped with one water closet for every 10 people using the building. The same ratio applies to lavatories. There must be one bathtub or shower for every 8 people in the building. A drinking fountain must be installed for every 100 people in the building. And, there is the requirement that one service sink be installed in the building.

Since you are preparing for a master plumber's exam, you will have to learn this type of information. Once you have learned some of the more common rules that pertain to fixtures, you can spend your spare time going over the data on minimum fixture requirements. For now, we will begin your introduction into plumbing fixtures with a multiple-choice exam.

Multiple-choice exam

1) A device that is of a special class of plumbing fixtures and is intended to perform a special function is generally considered which of the following:

   A) A bidet
   B) A combination fixture
   C) A plumbing appliance
   D) None of the above

2) Fixtures that depend on one or more energized components, such as motors, controls, heating elements, or pressure or temperature sensing elements are usually called _____. Fill in the blank.

   A) Plumbing appliances
   B) Mechanized fixtures
   C) Complex fixtures
   D) None of the above

3) A _____ doesn't require additional water supply or add additional discharge load to a fixture or drainage system. Fill in the blank.

   A) Plumbing appliance
   B) Plumbing appurtenance
   C) Plumbing fixture
   D) None of the above

4) A receptacle or device that is either permanently or temporarily connected to the water distribution system of the premises and demands water from the system is called _____. Fill in the blank.

   A) A plumbing appurtenance
   B) A plumbing fixture
   C) Both A and B
   D) Neither A nor B

5) A receptacle or device that is either permanently or temporarily connected to a plumbing system that discharges waste water, liquid-borne waste materials, or sewage either directly or indirectly into the drainage system is called _____. Fill in the blank.

A) A plumbing appurtenance
B) A plumbing fixture
C) Both A and B
D) Neither A nor B

6) A receptacle or device that requires both a water supply connection and a discharge to a drainage system is called:

A) A plumbing appurtenance
B) A plumbing fixture
C) Both A and B
D) A plumbing conservation device

7) A water closet that is installed in such a way that it does not touch a floor is called:

A) A carrier water closet
B) A hanger water closet
C) A wall-hung water closet
D) None of the above

8) A bathtub that is equipped and fitted with a circulation piping system, pump, and similar appurtenances, and is so designed to accept, circulate, and discharge bathtub water is called a:

A) Spa
B) Whirlpool bathtub
C) Either A or B
D) Both A and B

9) Which of the following locations would be considered to have fixtures meant for public use:

A) Toilet rooms in schools
B) Toilet rooms in gymnasiums
C) Both A and B
D) Neither A nor B

10) Which of the following locations would be considered to have fixtures meant for public use:

A) Toilet rooms in hotels
B) Toilet rooms in apartments
C) Both A and B
D) Neither A nor B

11) Which of the following locations would be considered to have fixtures meant for public use:

A) Toilet rooms in railroad stations
B) Toilet rooms in bars
C) Both A and B
D) Neither A nor B

12) Which of the following locations would be considered to have fixtures meant for public use:

A) Toilet rooms in public buildings
B) Toilet rooms in residences
C) Both A and B
D) Neither A nor B

13) Which of the following locations would be considered to have fixtures meant for public use:

A) Toilet rooms in public comfort stations
B) Toilet rooms in apartments
C) Both A and B
D) Neither A nor B

14) Which of the following locations would be considered to have fixtures meant for private use:

A) Toilet rooms in public comfort stations
B) Toilet rooms in apartments
C) Both A and B
D) Neither A nor B

15) Which of the following locations would be considered to have fixtures meant for private use:

A) Toilet rooms in residences
B) Toilet rooms in apartments
C) Both A and B
D) Neither A nor B

16) Nonpublic toilet rooms in hotels and motels are considered to have fixtures intended for:

A) Private use
B) Public use
C) Communal use
D) Commercial use

17) All plumbing fixtures and drains that receive or discharge wastes or sewage are required to be connected to:

A) A storm drain
B) A gray-water drain
C) A sanitary drainage system
D) None of the above

18) Plumbing fixtures in buildings are often required in separate facilities for each sex of potential occupants of the building. Which of the following types of buildings are exempt from having separate facilities for males and females:

A) Residential properties
B) Business properties where 15 people, or less, are employed
C) Buildings with a total occupant load, including both employees and customers, of 15 or less, in which food or beverage is served for consumption within the structure.
D) All of the above

19) Under normal conditions, when fixtures are being supplied in separate facilities within a single building, the ratio of fixtures provided for females and males should be:

A) Evenly distributed with 50 percent for each sex
B) Distributed with 75 percent of the fixtures being allotted for females
C) Distributed with 75 percent of the fixtures being allotted for males
D) None of the above

20) In most cases, access to toilet facilities for employees must be:

A) Within 50 feet of the employees' work space
B) Accessible from the employees' regular working area
C) Both A and B
D) Neither A nor B

21) Toilet facilities for employees, in most cases, must be located not more than _____ above or below the employees' regular work area. Fill in the blank.

A) 1 story
B) 2 stories
C) 18 feet
D) None of the above

22) Toilet facilities for employees, in most cases, must be located in such a way that employees will not have to travel more than _____ feet to reach them. Fill in the blank.

A) 50 C) 200
B) 100 D) 500

23) Some types of buildings require toilet facilities for employees. The facilities may be separate or the same facilities used for public customers. Which of the following uses require such facilities:

A) Restaurants C) Both A and B
B) Nightclubs D) None of the above

24) Some types of buildings require toilet facilities for employees. The facilities may be separate or the same facilities used for public customers. Which of the following uses require such facilities:

A) Places of public assembly C) Both A and B
B) Mercantile occupancies D) None of the above

25) For this question, you must consider a specific set of circumstances. You have been asked to plumb a tenant space in a shopping mall. The tenant space contains 800 square feet. The travel distance from the main entrance of the tenant space to a central toilet area in the mall is less than 500 feet. The central toilet facilities are located less than one story above or below the tenant space. What toilet facilities are required in the tenant space for the employees?

A) A water closet and lavatory
B) A water closet, urinal, and lavatory
C) No facilities are required
D) A hand sink is all that is required

26) Customers, patrons, and visitors to certain types of buildings are entitled to toilet facilities. Which of the following types of building uses will trigger the need for such toilet facilities?

A) Restaurants C) Both A and B
B) Mercantile occupancies D) None of the above

27) Customers, patrons, and visitors to certain types of buildings are entitled to toilet facilities. Which of the following types of building uses will trigger the need for such toilet facilities?

A) Restaurants
B) Nightclubs
C) Both A and B
D) None of the above

28) Customers, patrons, and visitors to certain types of buildings are entitled to toilet facilities. Which of the following types of building uses will trigger the need for such toilet facilities?

A) Restaurants
B) Places of assembly
C) Both A and B
D) None of the above

29) Toilet rooms and bathing facilities containing fixtures for occupants of a building that is required to be accessible to physically disadvantaged persons must have at least _____ of each type accessible to and usable by a physically disadvantaged person. Fill in the blank.

A) 1
B) 2
C) 3
D) 4

30) Toilet rooms and bathing facilities containing fixtures for occupants of a building that is required to be accessible to physically disadvantaged persons must have at least _____ wheelchair-accessible compartments for a water closet accessible to and usable by a physically disadvantaged person. Fill in the blank.

A) 1
B) 2
C) 3
D) 4

31) When 6 or more water-closet compartments are provided in a toilet room or bathing facility, at least _____ ambulatory-accessible compartment(s) must be provided, in addition to the required wheelchair-accessible compartment. Fill in the blank.

A) 1
B) 2
C) 3
D) 4

32) Where drinking fountains are installed on each floor, in buildings required to be accessible to physically disadvantaged persons, _____ percent, but not less than one drinking fountain or water

cooler per floor shall be accessible to and usable by physically disadvantaged persons. Fill in the blank.

A) 25
B) 33⅓
C) 50
D) 75

33) When showers are installed in gymnasiums in conjunction with sports activities, in buildings required to be adapted to meet the needs of physically disadvantaged persons, a minimum of ____ shall be accessible to and usable by physically disabled persons. Fill in the blank.

A) 1 shower
B) 2 showers
C) 3 showers
D) 4 showers

34) The minimum net clear opening to a water closet compartment shall be:

A) 30 inches
B) 32 inches
C) 36 inches
D) 42 inches

The questions you are about to encounter pertain to plumbing fixtures that are deemed to be accessible. All of the following questions, until further notice, pertain to accessible fixtures as recognized by the plumbing code.

35) Doors installed on water-closet compartments must swing _____. Fill in the blank.

A) In, toward the toilet
B) Out, away from the toilet
C) Either A or B
D) Neither A nor B

36) The minimum clear width in front of a door to a water-closet compartment shall be:

A) 32 inches
B) 36 inches
C) 42 inches
D) 48 inches

37) Grab bars installed in toilet compartments must be mounted parallel to the floor and no more than _____ inches above the finished floor. Fill in the blank.

A) 32
B) 33
C) 35
D) 36

38) When a toilet compartment is built to be wheelchair accessible, the minimum width of the compartment is:

A) 36 inches
B) 42 inches
C) 48 inches
D) 60 inches

39) The minimum length of a wheelchair-accessible toilet compartment, when a wall-hung toilet is installed, is:

A) 42 inches
B) 48 inches
C) 56 inches
D) 60 inches

40) The minimum length of a wheelchair-accessible toilet compartment, when a floor-mounted toilet is installed, is:

A) 48 inches
B) 52 inches
C) 56 inches
D) 59 inches

41) Water closets installed in compartments that are wheelchair accessible must be installed so that there is a minimum of ____ inches from the center of the toilet to the nearest side wall. Fill in the blank.

A) 12
B) 15
C) 18
D) 20

42) The minimum length for a side grab bar in a wheelchair-accessible toilet enclosure is:

A) 36 inches
B) 42 inches
C) 48 inches
D) 60 inches

43) The minimum length for a rear grab bar in a wheelchair-accessible toilet enclosure is:

A) 36 inches
B) 42 inches
C) 48 inches
D) 60 inches

44) Grab bars mounted on the side walls of wheelchair-accessible toilet compartments must be installed so that their point of beginning from the rear wall does not exceed _____ inches. Fill in the blank.

A) 12
B) 16
C) 18
D) 20

45) Grab bars mounted on the side walls of wheelchair-accessible toilet compartments must be installed so that they extend _____ inches from the rear wall of the compartments. Fill in the blank.

A) 36
B) 42
C) 48
D) 52

46) Grab bars mounted on the rear walls of wheelchair-accessible toilet compartments must be installed so that they extend a minimum of _____ inches beyond the centerline of a water closet. Fill in the blank.

A) 18
B) 24
C) 36
D) 48

47) Grab bars mounted on the rear walls of wheelchair-accessible toilet compartments must be installed so that they are not more than ____ inches from the closest side wall. Fill in the blank.

A) 6
B) 12
C) 16
D) 18

48) When a lavatory is installed in a compartment with a wheelchair-accessible toilet, the edge of the lavatory must not be less than ____ inches from the center of the toilet. Fill in the blank.

A) 6
B) 12
C) 16
D) 18

49) The maximum width of an ambulatory-accessible toilet compartment is:

A) 36 inches
B) 42 inches
C) 48 inches
D) 60 inches

50) The minimum length of an ambulatory-accessible toilet compartment is:

A) 48 inches
B) 56 inches
C) 60 inches
D) 72 inches

51) Water closets installed in compartments that are ambulatory-accessible must be installed so that there is a minimum of ____ inches from the center of the toilet to the nearest side wall. Fill in the blank.

A) 12
B) 15
C) 18
D) 20

52) Ambulatory-accessible toilet compartments are required to have grab bars that have a minimum length of _____ inches mounted on both sides of the compartment. Fill in the blank.

A) 36
B) 42
C) 48
D) 56

53) The side grab bars in an ambulatory-accessible toilet compartment must be installed so that they are not more than _____ inches from the rear wall of the compartment. Fill in the blank.

A) 6
B) 12
C) 18
D) 24

54) The side grab bars in an ambulatory-accessible toilet compartment must be installed so that they extend at least _____ inches from the rear wall of the compartment. Fill in the blank.

A) 42
B) 44
C) 52
D) 54

55) The seat of a water closet must be at least ____ inches above the finished floor. Fill in the blank.

A) 14
B) 16
C) 18
D) 20

56) The seat of a water closet must be no more than _____ inches above the finished floor. Fill in the blank.

A) 14
B) 16
C) 18
D) 20

57) Which of the following types of water closet bowls are not acceptable:

A) Siphon-vortex
B) Siphon-wash
C) Blowout
D) None of the above

58) Which of the following types of water closet bowls are acceptable:

A) Siphon-vortex
B) Siphon-wash
C) Blowout
D) All of the above

59) Which of the following types of water closet bowls are acceptable:

A) Siphon-jet
B) Siphon-wash
C) Reverse-trap
D) All of the above

60) What is the minimum clear space required in front of a lavatory?

A) 18 inches
B) 24 inches
C) 30 inches
D) 32 inches

61) The top of a lavatory must not be more than _____ inches above a finished floor. Fill in the blank.

A) 30
B) 32
C) 35
D) 36

62) The top of a kitchen sink must not be more than _____ inches above the finished floor. Fill in the blank.

A) 32
B) 34
C) 35
D) 36

63) What is the minimum unobstructed knee clearance, in height, that must be present for a lavatory?

A) 24 inches
B) 26 inches
C) 29 inches
D) 32 inches

64) What is the minimum unobstructed knee clearance, in depth, that must be present for a kitchen sink?

A) 6 inches
B) 8 inches
C) 10 inches
D) 12 inches

65) What is the minimum unobstructed knee clearance, in depth, that must be present for a lavatory?

A) 6 inches
B) 8 inches
C) 10 inches
D) 12 inches

66) What is the minimum unobstructed toe clearance, in depth, that must be present for a lavatory?

A) 8 inches
B) 9 inches
C) 10 inches
D) 12 inches

67) What is the minimum unobstructed toe clearance, in height, that must be present for a lavatory?

A) 8 inches
B) 9 inches
C) 10 inches
D) 12 inches

68) How far back from the face of a lavatory may the controls for the lavatory faucet be located?

A) 18 inches
B) 20 inches
C) 24 inches
D) 25 inches

69) A lavatory faucet must open and close with a maximum force of _____ pounds. Fill in the blank.

A) 3
B) 5
C) 7
D) 9

70) The spout of a drinking fountain must not be more than _____ inches above the finished floor. Fill in the blank.

A) 30
B) 32
C) 36
D) 42

71) The flow of water in a drinking fountain may rise up to _____ inches above the basin. Fill in the blank.

A) 2
B) 3
C) 4
D) 5

72) How much clear space is required in front of a water cooler or drinking fountain?

A) 18 inches
B) 24 inches
C) 30 inches
D) 36 inches

73) How much clear space, measured in width, is required in front of a water cooler or drinking fountain, excluding exceptions?

A) 28 inches
B) 32 inches
C) 36 inches
D) 48 inches

74) How much clear space is required in front of a bathtub?

A) 18 inches
B) 24 inches
C) 30 inches
D) 36 inches

75) How much clear space is required in front of a shower?

A) 18 inches
B) 24 inches
C) 30 inches
D) 36 inches

76) How much clear space, measured in width, is required in front of a bathtub?

A) 36 inches
B) 42 inches
C) 48 inches
D) 50 inches

77) How much clear space, measured in width, is required in front of a shower?

A) 36 inches
B) 39 inches
C) 42 inches
D) 48 inches

78) When a bathtub is not accessible from the side, the clear space required in front of the bathtub is:

A) 36 inches
B) 42 inches
C) 48 inches
D) 52 inches

79) When a bathtub is not accessible from the side, the clear space, measured in width, required in front of the bathtub is:

A) 36 inches
B) 42 inches
C) 48 inches
D) 52 inches

80) When a shower is not accessible from the side, the clear space required in front of the bathtub is:

A) 36 inches
B) 42 inches
C) 48 inches
D) 52 inches

81) When a shower is not accessible from the side, the clear space, measured in width, required in front of the bathtub is:

A) 36 inches
B) 42 inches
C) 48 inches
D) 52 inches

82) Bathtub enclosures shall have which of the following types of seats:

A) Built-in seats
B) Seats at the ends of the tubs
C) Detachable seats
D) Any of the above

83) How many grab bars are required on the rear wall of a bathtub enclosure?

A) 1
B) 2
C) 3
D) None of the above

84) When grab bars are installed on the rear walls of bathtub enclosures, the bars must be at least _____ inches in length. Fill in the blank.

A) 18
B) 20
C) 24
D) 30

85) When grab bars are installed on the rear walls of bathtub enclosures, the top grab bar must be at least _____ inches above the finished floor. Fill in the blank.

A) 24
B) 28
C) 30
D) 33

86) When grab bars are installed on the rear walls of bathtub enclosures, the top grab bar must be no more than _____ inches above the finished floor. Fill in the blank.

A) 28
B) 30
C) 34
D) 36

87) When grab bars are installed on the rear walls of bathtub enclosures, the bottom grab bar must be mounted _____ inches above the flood-level rim of the fixture. Fill in the blank.

A) 6
B) 9
C) 12
D) 16

88) The grab bar near the faucet in a bathtub enclosure must be at least _____ inches long. Fill in the blank.

A) 18
B) 20
C) 22
D) 24

89) What is the minimum width of a shower enclosure?

A) 30 inches
B) 32 inches
C) 48 inches
D) 60 inches

90) What is the minimum depth of a shower enclosure?

A) 30 inches
B) 32 inches
C) 48 inches
D) 60 inches

91) Wheelchair-accessible shower enclosures must not have _____. Fill in the blank.

A) Pressure-balanced valves
B) Hand-held showers
C) Thresholds
D) Grab bars

92) When a shower valve is installed in a wheelchair-accessible shower enclosure, what is the minimum distance above a finished floor that the valve can be installed?

A) 33 inches
B) 36 inches
C) 38 inches
D) 42 inches

93) When a shower valve is installed in a wheelchair-accessible shower enclosure, what is the maximum distance above a finished floor that the valve can be installed?

A) 36 inches
B) 42 inches
C) 44 inches
D) 48 inches

94) How many grab bars are required in a shower enclosure?

A) 1
B) 2
C) 3
D) 4

95) What is the minimum distance above a finished floor that a grab bar in a shower enclosure can be mounted?

A) 32 inches
B) 33 inches
C) 34 inches
D) 36 inches

96) What is the maximum distance above a finished floor that a grab bar in a shower enclosure can be mounted?

A) 32 inches
B) 33 inches
C) 34 inches
D) 36 inches

97) When a shower enclosure is square, what is the minimum size requirement for the enclosure?

A) 32 inches square
B) 34 inches square
C) 36 inches square
D) 42 inches square

98) What is the maximum length of a shower seat in a square shower enclosure?

A) 16 inches
B) 18 inches
C) 20 inches
D) 24 inches

99) What is the minimum distance above a finished floor that a shower seat, in a square shower enclosure, can be mounted?

A) 15 inches
B) 17 inches
C) 19 inches
D) 20 inches

100) What is the maximum distance above a finished floor that a shower seat, in a square shower enclosure, can be mounted?

A) 15 inches
B) 17 inches
C) 19 inches
D) 20 inches

101) Showers and bathtubs are required to have hand-held showers when used as accessible fixtures. How long does the hose on the hand-held shower have to be?

A) 30 inches
B) 36 inches
C) 48 inches
D) 60 inches

102) What is the minimum diameter or width of a grab bar?

A) 1 inch
B) 1¼ inches
C) 1½ inches
D) 2 inches

103) What is the maximum diameter or width of a grab bar?

A) 1 inch
B) 1¼ inches
C) 1½ inches
D) 2 inches

This concludes the section on accessible plumbing fixtures, but we are far from done with the testing on other fixtures. The remainder of this exam will pertain to all plumbing fixtures, unless otherwise noted.

104) What is the minimum distance of clear space required from the center of a water closet to the closest wall?

A) 12 inches
B) 15 inches
C) 18 inches
D) 30 inches

105) What is the minimum distance of clear space required from the center of a lavatory to the closest wall?

A) 12 inches
B) 15 inches
C) 18 inches
D) 30 inches

106) What is the minimum distance of clear space required from the center of a bidet to the closest wall?

A) 12 inches
B) 15 inches
C) 18 inches
D) 30 inches

107) How much clear space is required in front of a water closet?

A) 12 inches
B) 15 inches
C) 18 inches
D) 30 inches

108) How much clear space is required in front of a lavatory?

A) 12 inches
B) 15 inches
C) 18 inches
D) 30 inches

109) How much clear space is required in front of a bidet?

A) 12 inches
B) 15 inches
C) 18 inches
D) 30 inches

110) When water closets are installed side by side, what is the minimum distance required between the centerlines of the adjacent fixtures?

A) 12 inches
B) 15 inches
C) 18 inches
D) 30 inches

111) When lavatories are installed side by side, what is the minimum distance required between the centerlines of the adjacent fixtures?

A) 12 inches
B) 15 inches
C) 18 inches
D) 30 inches

112) When bidets are installed side by side, what is the minimum distance required between the centerlines of the adjacent fixtures?

A) 12 inches
B) 15 inches
C) 18 inches
D) 30 inches

113) What is the minimum distance of clear space required from the center of a urinal to the closest wall?

A) 12 inches
B) 15 inches
C) 18 inches
D) 30 inches

114) Fixtures equipped with concealed slip-joint connections, such as a tub waste and overflow, must be provided with _____. Fill in the blank.

A) An access panel
B) A utility space
C) Either A or B
D) None of the above

115) Fixtures equipped with concealed solvent-cemented joint connections, such as a tub waste and overflow, must be provided with _____. Fill in the blank.

A) An access panel
B) A utility space
C) Either A or B
D) None of the above

116) Water supplies to automatic clothes washers must be protected against back flow by a back-flow preventer or _____. Fill in the blank.

A) An airbreak
B) An airgap
C) Either A or B
D) Neither A nor B

117) The waste discharge from an automatic washer must be through _____. Fill in the blank.

A) An airbreak
B) An airgap
C) Either A or B
D) Neither A nor B

118) The minimum size allowed for a waste and overflow for a bathtub is:

A) 1¼ inches
B) 1½ inches
C) 2 inches
D) 2½ inches

119) Water supplies to bidets must be protected against back flow by a back-flow preventer or _____. Fill in the blank.

A) An airbreak
B) An airgap
C) Either A or B
D) Neither A nor B

120) Water supplies to dishwashers must be protected against back flow by a back-flow preventer or _____. Fill in the blank.

A) An airbreak
B) An airgap
C) Either A or B
D) Neither A nor B

121) Which of the following is a suitable place to accept the discharge from a dishwasher:

A) A trap
B) A trapped fixture
C) A food waste grinder
D) All of the above

122) Which of the following is a suitable place to accept the discharge from a dishwasher:

A) A trap
B) A trapped fixture
C) A tailpiece of a kitchen sink
D) All of the above

123) Drinking fountains may not be installed in which of the following locations:

A) Private bathrooms
B) Public bathrooms
C) Both A and B
D) Neither A nor B

124) Emergency showers must be provided with a supply of _____. Fill in the blank.

A) Cold water
B) Hot water
C) Both A and B
D) None of the above

125) Which of the following is not required for an emergency shower:

A) A hot water supply
B) A waste connection
C) Both A and B
D) None of the above

126) Which of the following is not required for an emergency eyewash station:

A) A hot water supply
B) A waste connection
C) Both A and B
D) None of the above

127) Floor drains must have a minimum size of:

A) 1½ inches
B) 2 inches
C) 3 inches
D) 4 inches

128) The strainers that cover the traps of floor drains must be _____. Fill in the blank.

A) Constructed of brass
B) Constructed of cast iron
C) Removable
D) Both A and C

129) The minimum size for a drain that accepts the waste from a domestic food waste grinder is:

A) 1¼ inches
B) 1½ inches
C) 2 inches
D) 3 inches

130) The minimum size for a drain that accepts the waste from a commercial food waste grinder is:

A) 1¼ inches
B) 1½ inches
C) 2 inches
D) 3 inches

131) Commercial dishwashers must _____. Fill in the blank.

A) Discharge through the tailpiece of a sink
B) Be trapped and connected to the drainage system separately from any other fixture or sink compartments
C) Discharge, by way of an airgap, into a deep sink bowl
D) None of the above

132) Food waste grinders must:

A) Discharge through a grease interceptor
B) Discharge through a grease interceptor when the grinder is being used for commercial purposes
C) Not discharge into a grease interceptor
D) None of the above

133) Water supplies to garbage can washers must be protected against back flow by a back-flow preventer or _____. Fill in the blank.

A) An airbreak
B) An airgap
C) Either A or B
D) Neither A nor B

134) Garbage can washers must _____. Fill in the blank.

A) Discharge through a sand interceptor
B) Be trapped separately
C) Both A and B
D) Either A or B

135) Receptors that receive waste from garbage can washers must _____. Fill in the blank.

A) Discharge through a sand interceptor
B) Be equipped with removable baskets or strainers
C) Both A and B
D) Either A or B

136) The minimum size of a waste outlet for a laundry tray is:

A) 1¼ inches
B) 1½ inches
C) 2 inches
D) 3 inches

137) Each compartment of a laundry tray must be equipped with _____. Fill in the blank.

A) A crossbar to restrict the clear opening of the waste outlet
B) A strainer to restrict the clear opening of the waste outlet
C) Both A and B
D) Either A or B

138) What effect does a food waste grinder have on the fixture-unit load of a sanitary drainage system?

A) It increases it
B) It increases it by 3 fixture units
C) It has no effect on the fixture-unit load
D) Both A and B

139) The minimum size of a waste outlet for a lavatory is:

A) 1¼ inches
B) 1½ inches
C) 2 inches
D) 3 inches

140) Which of the following is suitable for restricting the clear opening of a waste outlet in a lavatory:

A) A strainer
B) A pop-up plug
C) A cross bar
D) Any of the above

141) The minimum size for a waste outlet that serves a shower that is not part of a tub-shower combination is:

A) 1¼ inches
B) 1½ inches
C) 2 inches
D) 3 inches

142) What is the minimum size of the diameter of a shower strainer?

A) 1½ inches
B) 2 inches
C) 3 inches
D) 4 inches

143) What is the minimum size of the openings required in strainers for shower drains?

A) ¼ inch
B) ½ inch
C) ¾ inch
D) 1 inch

144) Shower compartments must contain a minimum of _____ square inches of interior cross-sectional area. Fill in the blank.

A) 600
B) 900
C) 1,100
D) 1,500

145) What is the minimum dimension of a shower when measured from the finished interior dimension of the compartment, exclusive of the fixture valves, shower heads, soap dishes, and safety grab bars?

A) 30 inches
B) 32 inches
C) 34 inches
D) 36 inches

146) When a bathtub is equipped with a shower head, the waterproof enclosure for the tub walls must extend at least _____ feet above the room floor level, and not less than _____ inches when

measured from the compartment floor at the drain. Fill in the blanks.

A) 6 feet and 72 inches
B) 5 feet and 60 inches
C) 6 feet and 70 inches
D) 6½ feet and 72 inches

147) When shower pans are formed from pan material, the sides of such pans must turn up on all sides by at least _____ inches. Fill in the blank.

A) 2
B) 2½
C) 3
D) 4

148) The minimum size of a waste outlet for a sink is:

A) 1¼ inches
B) 1½ inches
C) 2 inches
D) 3 inches

149) When sinks are equipped with food waste grinders, the sink's waste opening must have a minimum diameter of _____ inches. Fill in the blank.

A) 1½
B) 2
C) 3
D) 3½

150) When calculating the required fixtures for a specific building, it is often acceptable to substitute some urinals for some water closets. However, in no case can urinals be substituted for more than _____ percent of the required water closets. Fill in the blank.

A) 25
B) 40
C) 50
D) 60

151) Urinals provided for public use or employee toilet facilities must be of which of the following types:

A) Blowout urinals
B) Siphon-jet urinals
C) Wash-out urinals
D) Any of the above

152) Urinals provided for public use or employee toilet facilities _____ be equipped with strainers. Fill in the blank.

A) Must
B) Must not
C) May
D) Can

153) Urinals provided for public use or employee toilet facilities _____ have visible water trap seals. Fill in the blank.

A) Must
B) Must not
C) May
D) Can

154) Water closets installed for use by the public or by employees must be of an _____ type. Fill in the blank.

A) Flush-tank
B) Round-front
C) Elongated
D) None of the above

155) Water closets installed for use by the public or by employees must have a minimum water surface area of 8 inches by _____ inches. Fill in the blank.

A) 6
B) 7
C) 8
D) 9

156) Water closets installed for use by the public or by employees must have _____. Fill in the blank.

A) Black seats
B) Hinged round-front seats
C) Hinged closed-front seats
D) Hinged open-front seats

157) Water closets that depend on trap siphonage to discharge the fixture contents to the drainage system must be provided with _____. Fill in the blank.

A) A flushometer valve
B) A flush tank
C) Either A or B
D) Both A and B

158) Urinals that depend on trap siphonage to discharge the fixture contents to the drainage system must be provided with _____. Fill in the blank.

A) A flushometer valve
B) A flush tank
C) Either A or B
D) Both A and B

159) Clinical sinks that depend on trap siphonage to discharge the fixture contents to the drainage system must be provided with _____. Fill in the blank.

A) A flushometer valve
B) A flush tank
C) Either A or B
D) Neither A nor B

160) Flushing devices are not allowed to serve more than _____. Fill in the blank.

A) 1 fixture
B) 2 fixtures
C) 2 fixtures of the same type
D) None of the above

161) When installing a whirlpool tub, a plumber must be sure that the pump for the whirlpool is _____ the weir of the fixture trap. Fill in the blank.

A) Below
B) Above
C) On the same level as
D) None of the above

162) Pumps used with whirlpool tubs must be _____. Fill in the blank.

A) Accessible
B) Readily accessible
C) At least ½ HP
D) None of the above

163) A device used to control or guide the flow of water into, or convey water from, a fixture is called a _____. Fill in the blank.

A) Guide tube
B) Fixture control
C) Fixture fitting
D) None of the above

We are finally finished with the multiple-choice section of this chapter. I apologize for the long length of this exam, but the subject of plumbing fixtures is a complicated one. It is also a very important aspect of the licensing exam. As you can tell by the volume of material covered in this exam, you have a lot to study on the subject of fixtures. If you're still with me at this point, you deserve a break. Before we move on to the true-false tests for the various regions, take a break, rest your eyes, and walk around. Then move on to the true-false section that pertains to your region.

True-false test for zone 1

1) The bowls of water closets installed for public use must be elongated.

True False

2) Special-use materials are sometimes made of soapstone.

True False

3) Stainless steel, with a gauge of 16, may not be used as a sink material for sinks in kitchens of restaurants.

True False

4) A 12-inch access panel must be provided wherever concealed slip-joint connections exist.

True False

5) When exposed or accessible, 20-gauge drawn brass may be used as a material for continuous wastes.

True False

6) The minimum size of a drain that serves a bathtub is 2 inches.

True False

7) Dishwasher drains may not be connected to a sink tailpiece on the discharge side of a food waste disposal unit.

True False

8) Back-flow protection is not required for ornamental ponds.

True False

9) When urinals are installed side by side, there should be a minimum of 30 inches from the center of one urinal to the next closest urinal.

True False

10) Floor drains are not considered plumbing fixtures.

True False

11) The threshold of a shower must be at least 2 inches lower than the sides and rear wall of the base.

True False

12) All shower compartments must have a minimum finished interior space of 1024 square inches.

True False

13) Shower receptors cannot be built directly on the ground.

True False

14) All whirlpool tubs must be equipped with a removable panel.

True False

15) The circulation pumps used with whirlpool tubs must be installed above the crown weir of the trap.

True False

True-false test for zone 2

1) The minimum size for a lavatory drain is 1½ inches.

True False

2) Back-flow protection must be provided for all plumbing fixtures.

True False

3) The minimum size for a sink drain is 1¼ inches.

True False

4) A drinking fountain must be installed in all public toilet rooms that are accessible to individuals with physical restrictions.

True False

5) The strainers installed on floor drains must be removable.

True False

6) The minimum drain size for a commercial food waste grinder is 3 inches.

True False

7) The minimum drain size for a floor drain is 2 inches.

True False

8) Dishwashers, other than those installed in private living quarters, must be provided with a supply of hot water with a temperature of 180 degrees Fahrenheit.

True False

9) All dishwashers must be connected directly to a sanitary drainage system.

True False

10) Domestic food waste grinders may be equipped with a hand-operated water supply.

True False

True-false test for zone 3

1) Dormitories are required to have a minimum of 1 drinking fountain for each 100 people using the facility.

True False

2) All residential dwellings are required to have at least one hook-up for an automatic clothes washing machine.

True False

3) Buildings used as day nurseries are required to have a minimum of one water closet for every 15 people using the facility.

True False

4) Up to 3 garbage can washers may discharge into a common trap, so long as the trap is not more than 30 inches from a vent.

True False

5) Drinking fountains may not be installed in private bathrooms.

True False

6) The waste connection for an automatic clothes washer must enter the drainage system through an airgap.

True False

7) The drain inlets of floor drains must be readily accessible.

True False

8) A 2-inch drain is required when a domestic food waste grinder is installed.

True False

9) Emergency showers are not required to be connected to a sanitary drainage system.

True False

10) Receptors that receive the discharge from garbage can washers are required to be equipped with removable strainers.

True False

You have finished the tests on the subject of plumbing fixtures. Check your answers with the correct answers at the end of the chapter. Allow ample time to go over the questions and answers carefully. As you can tell by the length of this chapter, questions on plumbing fixtures might make up a large percentage of the licensing exam. When you are comfortable with your knowledge of fixtures, move on to the next chapter, where you will learn about plumbing for health care and institutional facilities.

Answers to multiple-choice exam

1) C
2) A
3) B
4) B
5) B
6) B
7) C
8) B
9) C
10) A
11) C
12) A
13) A
14) B
15) C
16) A
17) C
18) D
19) A
20) B
21) A
22) D
23) C
24) C
25) C
26) C
27) C
28) C
29) A
30) A
31) A
32) C
33) A
34) B
35) B
36) D
37) D
38) D
39) C
40) D
41) C
42) B
43) A
44) A
45) D
46) B
47) A
48) D

49) A
50) C
51) C
52) B
53) B
54) D
55) B
56) D
57) D
58) D
59) D
60) C
61) C
62) B
63) C
64) B
65) B
66) B
67) B
68) D
69) B
70) C
71) C
72) C
73) D
74) C
75) C
76) C
77) D
78) C
79) C
80) C
81) C
82) D
83) B
84) C
85) D
86) D
87) B
88) D
89) D
90) A
91) C
92) C
93) D
94) C
95) B
96) D
97) C
98) A
99) B
100) C
101) D
102) B
103) C
104) B
105) B
106) B
107) C
108) C
109) C
110) D
111) D
112) D
113) B
114) C
115) D
116) B
117) A
118) B
119) B
120) B
121) D
122) D
123) C
124) A
125) C
126) C
127) B
128) C
129) B
130) C
131) B
132) C
133) B
134) B
135) B
136) B
137) D
138) C
139) A
140) D
141) C
142) C
143) A
144) B
145) A
146) C
147) A
148) B
149) D
150) C
151) D
152) B
153) A
154) C
155) B
156) D
157) C
158) C
159) C
160) A
161) B
162) A
163) C

Answers to true-false test for zone 1

1) T
2) T
3) F
4) T
5) T
6) F
7) T
8) F
9) F
10) F
11) F
12) T
13) F
14) T
15) T

Answers to true-false test for zone 2

1) F	5) T	8) T
2) T	6) F	9) F
3) F	7) T	10) T
4) F		

Answers to true-false test for zone 3

1) T	5) T	8) F
2) T	6) F	9) T
3) T	7) T	10) T
4) F		

CHAPTER 17

Health care and institutional plumbing

NOT all plumbers work with health care and institutional plumbing, but everyone taking the plumber's licensing exam should know the basics of this specialized type of plumbing. While the plumbing code that pertains to health care and institutional facilities is not extensive, you might see several questions on the licensing exam that have been extracted from this section of the code.

When you read the plumbing code in the area of health care plumbing, you might see words that you are not accustomed to using. Fixtures will be mentioned that many plumbers have never seen or installed. The world of health care plumbing is very different from that of routine, residential and commercial plumbing.

I have talked previously about how the questions on licensing exams can be tricky because of the way they are worded. When you begin to read questions written about bedpan washers, aspirators, sterilizers, local vent stacks, and other plumbing devices used in medical facilities, the level of confusion can rise dramatically.

If you are not familiar with the section of the plumbing code that pertains to health care facilities, read through it before you take the tests in this chapter. Get acquainted with the words, terms, and phrases used. When you're ready, take the multiple-choice exam that follows.

Multiple-choice exam

1) Which of the following devices is supplied with water, or some other fluid, under positive pressure that passes through an integral orifice or constriction, causing a vacuum?

A) Sterilizer vent
B) Sterilizer
C) Aspirator
D) None of the above

2) Which of the following is a bedpan washer hose supplied with:

A) Cold water
B) Hot water
C) A and B
D) A vent

3) Which of the following types of traps are required for use with clinical sinks?

A) P-traps
B) Drum traps
C) Crown-vented traps
D) Integral traps

4) By which of the following means may the contents of a clinical sink be removed?

A) Raising a pop-up plug
B) Pulling a basket strainer
C) Siphonic action
D) None of the above

5) By which of the following means may the contents of a clinical sink be removed?

A) Raising a pop-up plug
B) Operating a waste-and-overflow device
C) Blowout action
D) None of the above

6) A clinical sink is required to have _____. Fill in the blank.

A) A stainless steel body
B) An S-trap
C) A visible water trap seal
D) None of the above

7) A clinical sink is required to have _____. Fill in the blank.

A) A stainless steel body
B) A flushing rim
C) Both A and B
D) None of the above

8) Service sinks may not be used for the disposal of _____. Fill in the blank.

A) Urine
B) Fecal matter
C) Human waste
D) All of the above

9) Which of the following is prohibited from being installed in a soiled laundry room?

A) Clinical sinks
B) Service sinks
C) Ice machines
D) All of the above

10) Which of the following is prohibited from being installed in a soiled laundry room?

A) Clinical sinks
B) Service sinks
C) Storage facilities for ice
D) All of the above

11) Clinical sinks that are installed in soiled laundry rooms may not be considered as substitutes for _____. Fill in the blank with the answer most appropriate to the code.

A) Service sinks
B) Local vent stacks
C) Laundry tubs
D) Lavatories

12) The condensate drainage from the steam supply of a sterilizer unit must be discharged _____. Fill in the blank.

A) Into a separate drainage system
B) By gravity
C) Into high-temperature piping
D) All of the above

13) Sterilizer piping must be _____. Fill in the blank.

A) Accessible
B) Readily accessible
C) Visibly accessible
D) Somewhat accessible

14) The piping for sterilizing equipment must be drained to prevent _____ from reaching the sterilizer. Fill in the blank with the answer that is most representative of the plumbing code.

A) Dirt
B) Sediment
C) Moisture
D) None of the above

15) Nonpressure sterilizers must be equipped to _____ vapors within the vessel. Fill in the blank with the answer which is most representative of the plumbing code.

A) Cool
B) Expel
C) Confine
D) None of the above

16) Pressure sterilizers must be equipped to do what with steam vapors? Fill in the blank with the answer that is most representative of the plumbing code.

A) Confine
B) Condense
C) Cool
D) Both B and C

17) The pipes and traps installed in mental health centers must be _____. Fill in the blank.

A) Insulated
B) Readily accessible
C) Concealed
D) Cool to the touch

18) Fixtures installed in mental health centers must be _____. Fill in the blank.

A) Insured
B) Nonbreakable
C) Bolted through the walls
D) All of the above

19) Which of the following is another name for a clinical sink?

A) Scrub sink
B) Service sink
C) Bedpan washer
D) None of the above

20) Which of the following types of fixtures are required to connect to a local vent?

A) Lavatory
B) Water closet
C) Kitchen sink
D) Bedpan washer

21) Which of the following must discharge to the drainage system through an indirect waste?

A) Sterilizers
B) Steamers
C) Condensers
D) All of the above

22) Which of the following must discharge to the drainage system through an airgap?

A) Sterilizers
B) Steamers
C) Condensers
D) All of the above

23) An indirect waste pipe for a bedpan steamer must be _____. Fill in the blank.

A) Made of cast iron
B) Made of copper
C) Trapped
D) None of the above

24) When two sterilizers discharge into an individual receptor, the distance between the receptor and the sterilizers must not exceed _____ feet. Fill in the blank.

A) 3
B) 5
C) 8
D) 12

25) Vacuum-system station receptacles must be provided with _____. Fill in the blank.

A) Access
B) Ready access
C) Glass enclosures
D) Both A and C

26) Vacuum-system station receptacles must be _____. Fill in the blank.

A) Built into cabinets
B) Built into recesses
C) Either A or B
D) Neither A nor B

27) Vacuum-system station receptacles must be _____. Fill in the blank.

A) Accessible
B) Readily accessible
C) Visible
D) Both B and C

28) Overflow prevention devices must be installed at each station outlet when bottle-type vacuum systems are used for collecting, removing, and disposing of _____. Fill in the blank.

A) Blood
B) Pus
C) Other fluids
D) All of the above

29) All central fluid suction systems must provide _____. Fill in the blank.

A) Continuous service
B) Intermittent service
C) Regular service
D) None of the above

30) When installed in hospitals, central fluid suction systems must be _____. Fill in the blank.

A) Installed in duplicate
B) Connected to the emergency power system
C) Shut-down at least once each week
D) None of the above

31) Exhausts from a vacuum pump that serves a vacuum fluid suction system must discharge _____. Fill in the blank.

A) Into a sanitary drainage system
B) Into open air above a roof
C) Into a sterilizer
D) Either A or B

32) The velocity of air flow in a central vacuum fluid-suction system must not be less than _____ feet per minute. Fill in the blank.

A) 2500
B) 3500
C) 5000
D) 7500

33) How many types of apparatus may be served by a single local vent?

A) 1
B) 2
C) 3
D) 4

34) What is the minimum diameter for a local vent that serves a bedpan washer?

A) 1¼ inches
B) 1½ inches
C) 2 inches
D) 3 inches

35) To which of the following may a local vent connect?

A) A sanitary drainage system
B) A sanitary venting system
C) A dryer vent
D) None of the above

36) Which of the following types of materials is suitable for use in central vacuum fluid-suction systems?

A) Corrosion-resistant materials
B) Corrosive materials
C) Either A or B
D) Neither A nor B

37) What is the minimum diameter of a branch for one outlet in a central vacuum fluid-suction system?

A) ¼ inch
B) ½ inch
C) ¾ inch
D) 1 inch

38) What is the minimum diameter of a main for one outlet in a central vacuum fluid-suction system?

A) ¼ inch
B) ½ inch
C) ¾ inch
D) 1 inch

39) On which of the following sections of piping for a central vacuum fluid-suction system is a cleanout required?

A) Branches
B) Mains
C) Both A and B
D) None of the above

40) When a cleanout is installed in a central vacuum fluid-suction system, what must be provided for the cleanout?

A) Raised-head plug
B) Recessed-head plug
C) Access
D) Raised-head cap

41) When a _____ is serving a single bedpan washer, it is permitted to drain into the fixture being served. Fill in the blank.

A) Local vent
B) Common vent
C) Wet vent
D) None of the above

42) How many local vents may be connected to a 2-inch local vent stack?

A) 1
B) 2
C) 3
D) 4

43) How many local vents may be connected to a 3-inch local vent stack?

A) 2
B) 4
C) 6
D) 8

44) How many local vents may be connected to a 4-inch local vent stack?

A) 5
B) 6
C) 8
D) 12

45) Which of the following types of fittings can be used to connect a local vent from a bedpan washer to a local vent stack?

A) Tee
B) Tee-wye
C) Either A or B
D) Neither A nor B

46) When a local vent from a bedpan washer is connected to a local vent stack, in which of the following positions must the connecting fitting be?

A) Horizontal
B) Upright
C) Upside down
D) Either A or B

47) What is the minimum size of a trap and waste for a local vent stack?

A) 1½ inches
B) 2 inches
C) The trap and waste must be the same size as the local vent stack
D) The trap and waste must be at least one-half the size of the local vent stack, but may never be smaller than 1½ inches

48) When is a local vent stack not required to be drained by means of a trapped-and-vented waste connection to the sanitary drainage system?

A) When it is installed in a dental office
B) When it is installed in a nursing home
C) When it serves only one bedpan washer
D) None of the above

49) What is the minimum diameter for a water supply pipe that is taken from the flush supply of a bedpan washer on the discharge or fixture side of the vacuum breaker?

A) ¼ inch
B) ½ inch
C) ¾ inch
D) 1 inch

50) For which of the following reasons is a water supply required to be connected to a local vent stack?

A) Cleansing
B) Drain trap seal maintenance
C) Both A and B
D) Neither A nor B

51) What is the required minimum depth of a water seal when a water supply is connected to a local vent stack?

A) 2 inches
B) 3 inches
C) 4 inches
D) 6 inches

52) When vents are installed for use with sterilizers, they must be _____. Fill in the blank.

A) Made of plastic
B) At least 2 inches in diameter
C) Accessible
D) Labeled

53) What is the minimum size of a sterilizer vent that serves a bedpan steamer?

A) 1¼ inches
B) 1½ inches
C) 2 inches
D) 3 inches

54) What is the minimum size of a pressure sterilizer vent?

A) 1¼ inches
B) 1½ inches
C) 2 inches
D) 2½ inches

55) What is the minimum size of a sterilizer vent stack that serves an instrument washer sterilizer?

A) 1¼ inches
B) 1½ inches
C) 2 inches
D) 3 inches

56) How many instrument washer sterilizers may be installed on a 2-inch vent stack?

A) 2
B) 3
C) 4
D) 6

57) How many instrument washer sterilizers may be installed on a 3-inch vent stack?

A) 2
B) 3
C) 4
D) 6

58) How many water service pipes are hospitals required to have?

A) 1
B) 2
C) 3
D) 4

59) The water service system for a hospital must be designed to provide _____. Fill in the blank.

A) 120 pounds per square inch of water pressure
B) 150 pounds per square inch of water pressure
C) Uninterrupted service
D) None of the above

60) How high above a fixture's flood-level rim must vacuum breakers be when installed on hospital fixtures?

A) 4 inches
B) 6 inches
C) 8 inches
D) 12 inches

61) The water supply for hospital fixtures must be protected against _____. Fill in the blank.

A) Back flow
B) Down flow
C) Up flow
D) None of the above

62) A condensate drain trap must discharge its contents through an _____. Fill in the blank.

A) Airgap
B) Airbreak
C) Either A or B
D) Neither A nor B

63) The cleaning, flushing, and resealing of a condensate drain trap seal must be done with _____. Fill in the blank.

A) Sterilized water
B) Cold water
C) A water supply
D) All of the above

64) How many vents from bedpan steamers or boiling-type sterilizers, which have a diameter of 1½ inches, can be connected to a 1½-inch sterilizer vent stack?

A) 0
B) 1
C) 2
D) 4

65) How many vents from bedpan steamers or boiling-type sterilizers, which have a diameter of 2 inches, can be connected to a 1½-inch sterilizer vent stack?

A) 0
B) 1
C) 2
D) 4

66) How many vents from bedpan steamers or boiling-type sterilizers, which have a diameter of 2 inches, can be connected to a 2-inch sterilizer vent stack?

A) 0
B) 1
C) 2
D) 4

67) How many vents from bedpan steamers or boiling-type sterilizers, which have a diameter of 1½ inches, can be connected to a 2-inch sterilizer vent stack?

A) 0
B) 1
C) 2
D) 4

68) How many vents from bedpan steamers or boiling-type sterilizers, which have a diameter of 1½ inches, can be connected to a 3-inch sterilizer vent stack?

A) 0
B) 1
C) 2
D) 4

69) How many vents from bedpan steamers or boiling-type sterilizers, which have a diameter of 2 inches, can be connected to a 3-inch sterilizer vent stack?

A) 0
B) 1
C) 2
D) 4

70) How many vents from bedpan steamers or boiling-type sterilizers, which have a diameter of 2 inches, can be connected to a 4-inch sterilizer vent stack?

A) 2
B) 4
C) 6
D) 8

71) How many vents from bedpan steamers or boiling-type sterilizers, which have a diameter of 1½ inches, can be connected to a 4-inch sterilizer vent stack?

A) 2
B) 4
C) 6
D) 8

72) How many vents from pressure sterilizers, which have a diameter of ¾ inch, can be connected to a 1½-inch sterilizer vent stack?

A) 0
B) 1
C) 2
D) 3

73) How many vents from pressure sterilizers, which have a diameter of 1 inch, can be connected to a 1½-inch sterilizer vent stack?

A) 1
B) 2
C) 3
D) 4

74) How many vents from pressure sterilizers, which have a diameter of 1¼ inches, can be connected to a 1½-inch sterilizer vent stack?

A) 1
B) 2
C) 3
D) 4

75) How many vents from pressure sterilizers, which have a diameter of 1½ inches, can be connected to a 1½-inch sterilizer vent stack?

A) 0
B) 2
C) 4
D) 6

76) How many vents from pressure sterilizers, which have a diameter of ¾ inch, can be connected to a 2-inch sterilizer vent stack?

A) 0
B) 2
C) 4
D) 6

77) How many vents from pressure sterilizers, which have a diameter of 1 inch, can be connected to a 2-inch sterilizer vent stack?

A) 0
B) 2
C) 3
D) 4

78) How many vents from pressure sterilizers, which have a diameter of 1¼ inches, can be connected to a 2-inch sterilizer vent stack?

A) 0
B) 2
C) 3
D) 4

79) How many vents from pressure sterilizers, which have a diameter of 1½ inches, can be connected to a 2-inch sterilizer vent stack?

A) 1
B) 2
C) 3
D) 4

80) How many vents from pressure sterilizers, which have a diameter of 1½ inches, can be connected to a 3-inch sterilizer vent stack?

A) 1
B) 2
C) 3
D) 4

81) How many vents from pressure sterilizers, which have a diameter of 1¼ inches, can be connected to a 3-inch sterilizer vent stack?

A) 2
B) 3
C) 5
D) 6

82) How many vents from pressure sterilizers, which have a diameter of 1 inch, can be connected to a 3-inch sterilizer vent stack?

A) 2
B) 4
C) 5
D) 7

83) How many vents from pressure sterilizers, which have a diameter of ¾ inch, can be connected to a 3-inch sterilizer vent stack?

A) 5
B) 10
C) 15
D) 18

84) Water sterilizers filled with water through directly connected piping must be equipped with _____. Fill in the blank.

A) A leakage diverter
B) A bleed line
C) Either A or B
D) Neither A nor B

85) Aspirators in laboratories must be protected against back flow with the use of _____. Fill in the blank.

A) An airgap
B) A vacuum breaker
C) Either A or B
D) An airbreak

86) Portable aspirators must be protected against back flow with the use of _____. Fill in the blank.

A) An airgap
B) A vacuum breaker
C) Either A or B
D) An airbreak

87) Vacuum system aspirators must be protected against back flow with the use of _____. Fill in the blank.

A) An airgap
B) A vacuum breaker
C) Either A or B
D) An airbreak

88) Bedpan washers must be protected against back flow with the use of _____. Fill in the blank.

A) An airgap
B) A vacuum breaker
C) Either A or B
D) An airbreak

89) Bedpan washer hoses must be protected against back flow with the use of _____. Fill in the blank.

A) An airgap
B) A vacuum breaker
C) Either A or B
D) An airbreak

90) Boiling-type sterilizers must be protected against back flow with the use of _____. Fill in the blank.

A) An airgap
B) A vacuum breaker
C) Either A or B
D) An airbreak

91) Exhaust condensers must be protected against back flow with the use of _____. Fill in the blank.

A) An airgap
B) A vacuum breaker
C) Either A or B
D) An airbreak

92) Flush floor drains must be protected against back flow with the use of _____. Fill in the blank.

A) An airgap
B) A vacuum breaker
C) Either A or B
D) An airbreak

93) Hose connections in health care and laboratory areas must be protected against back flow with the use of _____. Fill in the blank.

A) An airgap
B) A vacuum breaker
C) Either A or B
D) An airbreak

94) Pressure instrument washer sterilizers must be protected against back flow with the use of _____. Fill in the blank.

A) An airgap
B) A vacuum breaker
C) Either A or B
D) An airbreak

95) Pressure sterilizers must be protected against back flow with the use of _____. Fill in the blank.

A) An airgap
B) A vacuum breaker
C) Either A or B
D) An airbreak

96) Vacuum systems of the cleaning-type must be protected against back flow with the use of _____. Fill in the blank.

A) An airgap
B) A vacuum breaker
C) Either A or B
D) An airbreak

97) Vacuum systems of the fluid-suction type must be protected against back flow with the use of _____. Fill in the blank.

A) An airgap
B) A vacuum breaker
C) Either A or B
D) An airbreak

98) Vacuum breakers installed to protect bedpan washer hoses must be installed _____ feet above the floor. Fill in the blank.

A) 3
B) 4
C) 5
D) 6

99) An airgap used for a boiling-type sterilizer must not be less than _____ the effective opening of the water supply. Fill in the blank.

A) Half
B) Twice
C) Three times
D) None of the above

100) Vacuum breakers installed to protect hose connections in health care and laboratory areas must not be installed less than _____ above the finished floor. Fill in the blank.

A) 42 inches
B) 3 feet
C) 60 inches
D) 6 feet

This is the end of the multiple-choice exam. Obviously, there are a number of rules and regulations that pertain to health care and institutional plumbing. And, as you can see, the code applications are a bit different from what they are for most plumbing.

Unless you have studied closely the health care section of your code, I doubt that you did very well on this exam. Don't feel bad, even experienced plumbers would have trouble attaining a high score on this type of test. After you take the true-false test that applies to your region, devote some serious time to studying the requirements for health care and institutional plumbing. The subject will not involve a majority of the licensing exam, but there could be enough questions on this topic to influence your overall score. Even if you never plan to work in a hospital or related structure, learn this section of the code before your sitting for the licensing exam. Now, move onto the true-false questions and see how you do.

True-false test for zone 1

The code book for plumbers in zone 1 does not contain a special section dedicated to health care or institutional plumbing. The code book index shows a reference to special wastes and indirect wastes, but we have covered much of that material earlier in this book. For this reason, there is no true-false test for zone 1 on the issue of health care or institutional plumbing.

True-false test for zone 2

1) A table that is used for post-mortem examination of a body is called an autopsy table.

   True False

2) A bedpan hopper is the same thing as a clinical sink.

   True False

3) A bedpan unit is a metal table that is used to support bedpans.

   True False

4) A bedpan washing hose must be supplied with both hot and cold water.

   True False

5) A scrub sink is used by operating personnel to scrub their hands prior to an operating procedure.

   True False

6) Scrub sinks must be equipped with either knee-action mixing valves or by wrist or pedal controls.

   True False

7) A still is a device used to distill liquids.

True False

8) It is unlawful to descale the interior of water sterilizers, stills, or similar equipment while the equipment is still connected to the water or drainage system.

True False

9) Control valves installed in hospital corridors in such a fashion as to protrude from a wall or opening must be installed to avoid injury to individuals passing the protruding objects.

True False

10) Sterilizers must be provided with individual and separate indirect wastes, with airgaps of a size equal to that of the tailpiece from the fixture.

True False

11) Not more than 4 sterilizers may be installed as a battery assembly and allowed to share a common trap.

True False

12) The minimum trap seal for a trap serving a bedpan steamer shall be at least 3 inches in depth.

True False

13) Wastes from central fluid-suction vacuum systems must not be connected directly to a sanitary drainage system.

True False

14) The lowest point of a condensate riser does not have to be trapped if it discharges over an indirect waste sink.

True False

15) A running trap can be used to trap a condensate riser.

True False

16) No more than one type of apparatus may be served by a single local vent.

True False

17) The minimum diameter of a local vent for a bedpan washer is 1½ inches.

True False

18) It shall be unlawful for a local vent serving a single bedpan washer to drain into the fixture being served.

True False

19) The minimum diameter of a sterilizer vent serving a bedpan sterilizer is 1½ inches.

True False

20) The minimum diameter of a sterilizer vent serving a utensil sterilizer is 1½ inches.

True False

21) The minimum diameter of a sterilizer vent serving an instrument sterilizer is 2 inches.

True False

22) Sterilizer vent stacks must not be smaller than 2½ inches in diameter.

True False

23) Not more than 3 instrument washer-sterilizers shall be connected to a 2-inch vent.

True False

24) The water heating equipment used in a hospital must be capable of providing 6½ gallons of water, at 125 degrees Fahrenheit, per hour per bed for hospital fixtures.

True False

25) The water heating equipment used in a hospital must be capable of providing 4 gallons of water, at 180 degrees Fahrenheit, per hour per bed for hospital kitchen fixtures.

True False

26) The water heating equipment used in a hospital must be capable of providing 4½ gallons of water, at 180 degrees Fahrenheit, per hour per bed for hospital laundry fixtures.

True False

27) Hospitals with more than 2 stories are required to be plumbed with hot-water risers with circulators.

True False

28) A fluid-suction vacuum system can be protected from back flow with the use of either a vacuum breaker or an airgap.

True False

29) Boiling-type sterilizers can be protected from back flow with the use of either a vacuum breaker or an airgap.

True False

True-false test for zone 3

1) All devices, appurtenances, and appliances intended to serve some special purpose, such as sterilization, and that connect to either the water supply or drainage system, must be protected against back flow.

True False

2) Multiple installations of bedpan washers are prohibited.

True False

3) Steam supplies to sterilizers must be drained to prevent moisture from entering the sterilizer.

True False

4) Bedpan washers must be connected to local vents.

True False

5) Condensers must drain through an indirect waste through the use of an airbreak.

True False

6) Sterilizers must drain through an indirect waste through the use of an airbreak.

True False

7) Steamers must drain through an indirect waste through the use of an airgap.

True False

8) Vacuum system stations must be accessible.

True False

9) The bottom of a local vent serving bedpan washers, except when only one bedpan washer is served, must be trapped.

True False

10) All hospitals must have at least 2 water service pipes installed.

True False

Congratulations, you have completed all of the tests on health care plumbing. I suspect it has been a bit more complicated than what you were expecting. There is no question that the code that applies to work in hospitals and other health care facilities is complex.

The correct answers for the questions in the previous tests are located at the end of this chapter. Compare your answers with the correct answers and grade your tests. Don't feel bad if you didn't do very well. Accept the fact that health care plumbing requires some special studying, and then consult your local code book.

Professional plumbers who specialize in health care and institutional plumbing can make good money. If you learn to install and service health care plumbing quickly and in compliance with the local plumbing code, you can be in nearly constant demand.

Now, let's move on to chapter 18. The chapter is going to deal with plumbing inspections, maintenance, and the testing of plumbing systems. This is information all working plumbers must learn.

Answers to multiple-choice exam

1) C	26) C	51) B	76) D
2) C	27) D	52) C	77) C
3) D	28) D	53) B	78) B
4) C	29) A	54) D	79) A
5) C	30) B	55) C	80) C
6) C	31) B	56) A	81) C
7) B	32) C	57) C	82) D
8) D	33) A	58) B	83) C
9) C	34) C	59) C	84) C
10) C	35) D	60) B	85) B
11) A	36) A	61) A	86) B
12) B	37) B	62) A	87) B
13) A	38) D	63) C	88) B
14) C	39) C	64) B	89) B
15) C	40) C	65) A	90) A
16) D	41) A	66) B	91) B
17) C	42) C	67) C	92) B
18) C	43) C	68) D	93) B
19) C	44) D	69) C	94) B
20) D	45) C	70) B	95) B
21) D	46) B	71) D	96) C
22) D	47) C	72) D	97) C
23) C	48) C	73) B	98) C
24) C	49) A	74) A	99) B
25) B	50) C	75) A	100) D

Answers to true-false test for zone 2

1) T	11) F	21) F
2) T	12) T	22) T
3) F	13) F	23) F
4) T	14) F	24) T
5) T	15) T	25) T
6) T	16) T	26) T
7) T	17) F	27) F
8) T	18) F	28) T
9) T	19) T	29) F
10) F	20) F	

Answers to true-false test for zone 3

1) T	5) F	8) F
2) F	6) F	9) T
3) T	7) T	10) T
4) T		

CHAPTER 18

Inspecting, maintaining, and testing systems

HOW important are inspections, maintenance, and the testing of plumbing systems? If you're doing work that requires an approved inspection before it can be placed into service, inspections are an important part of plumbing. Not all plumbing work requires inspections, but much of it does. It pays to know what is expected of you from the local plumbing inspectors.

Maintaining a plumbing system is not really the responsibility of plumbers. This burden falls to property owners. Plumbers, however, are usually summoned to work on defective systems, so from that point of view, it has some bearing on the role of plumbers.

Testing is a safeguard for the plumber who installs work and the company for whom the plumber works. When plumbing is tested and inspected properly, much of the liability for which a plumber might be responsible is removed. Let me give you a quick example of what I mean.

Assume that you have been hired to install new plumbing in a remodeling job. Your work entails running new waste, vent, and water pipes from the basement of a home to a new bathroom that is being installed in converted attic space. Your pipes will be concealed behind the drywall, in one of the living room partitions. The job is a small one, but it does require the issuance of a permit and proper inspections by the local plumbing inspector.

When you are ready to start the job, the homeowners tell you that they have decided not to obtain permits for the work. They don't want their property taxes to be re-evaluated and raised. You call your master plumber and confer about doing the work without a permit. The master plumber hedges on the issue, but doesn't want to lose the job, so he authorizes you to do the work without a permit. Does this make what you're doing okay? No, but most of the burden, if you get caught, will fall on the master plumber. Since you want to keep your job, you do the work as instructed.

After you complete the plumbing installation, you turn the water back on and test your water pipes. There are no apparent leaks. Since you are in the rough-in stage, there are no fixtures to fill and drain, so you assume the drain, waste, and vent pipes will not leak. Your boss has told you not to bother testing these pipes, since the job is not being inspected, and plastic drains and vents rarely leak.

The homeowners have a drywall contractor repair the finished wall where your plumbing is installed, and the remodeling job is completed. When you set fixtures, you fill and flush all of the fixtures, and everything seems to work fine. The homeowners are happy, and you move on to another job. It is not until a few weeks later that trouble arises.

After using their new bathroom for a few weeks, the homeowners call your office. They are furious. It seems there is a plumbing leak in the recently installed work and it has damaged their living room ceiling, part of the living room wall, some expensive furniture, and an exotic, oriental rug. Your master plumber goes to visit the homeowners and inspects the damage. After removing some of the damaged ceiling, the master plumber discovers that one of the 3-inch fittings in the drain line for the toilet was never glued. Every time the toilet is flushed, a significant amount of water seeps out of the dry-fitted joint. This water has been leaking since the toilet was installed, with the water pooling on the living room ceiling.

The water that collected in the ceiling spread out and was absorbed by the drywall in the ceiling and wall. Ultimately, the volume of water was sufficient to come through the ceiling and damage expensive items in the living room. The homeowners not only want the leak fixed, they want restitution for the damages incurred.

When the homeowners threaten to sue the plumbing company, the owner and master plumber realize they are on very shaky ground. They did a job in violation of the plumbing code, knowing they were being negligent and doing wrong. This means the company's insurance will probably not pay for the damages. Even though the homeowners asked that the job be done without permits and code-enforcement inspections, they could easily deny this fact and press on in a legal battle.

It is easy to see how you, the master plumber, and the plumbing company are now in serious trouble. Could this have been avoided if the work had been done in compliance with the plumbing code? Yes, a proper test and inspection would have revealed the leak before it did any damage.

This example is not far from some true examples I've seen in the past. Many plumbers bootleg jobs to avoid the costs and hassles involved with permits and inspections. In most cases they don't get caught and nothing goes wrong. However, it only takes one major problem to ruin a plumbing career or company. If you think the testing and inspection of plumbing is not important, you had better spend some time reviewing the possibilities of what could happen if you don't play by the rules.

In the example above, the leak would have been found with a proper test. However, even if the leak didn't surface, the plumbing company would have been on much better legal ground if it had done the job in compliance with the plumbing code. At least then the insurance company would have to pay for the damages, since the work would have been inspected and approved by a code-enforcement officer.

This chapter is not going to be very long, and the rules and regulations we will cover are not difficult to learn or understand. The rules are few, but you will more than likely see some of them on the licensing exam. Let's turn our attention to the multiple-choice exam, and see how you do.

Multiple-choice exam

1) A plumbing system shall not be _____ until the system has been inspected and approved by the appropriate authorities. Fill in the blank.

A) Enclosed
B) Covered
C) Put into operation
D) All of the above

2) Code officials may require up to _____ hours of notice in order to schedule a plumbing inspection. Fill in the blank.

A) 8
B) 12
C) 24
D) 48

3) Who is responsible for testing the installation of new plumbing?

A) A journeyman plumber
B) A master plumber
C) The property owner
D) The permit holder

4) Who is responsible for the expense of testing a new plumbing system?

A) A journeyman plumber
B) A master plumber
C) The plumbing contractor installing the plumbing
D) The property owner

5) Which of the following may be used to test drain, waste, and vent pipes?

A) Air
B) Water
C) Either A or B
D) Neither A nor B

6) Which of the following may be used to test water pipes?

A) Air
B) Water
C) Either A or B
D) Neither A nor B

7) Which of the following elements of a drainage system are exempt from rough-in testing?

A) Outside rain leaders
B) Open-jointed drain tile
C) Both A and B
D) Neither A nor B

8) A plumbing official may require _____ to be removed in order to assure that test pressure is reached at all sections of piping. Fill in the blank.

A) Vent terminals
B) Cleanout plugs
C) Test gauges
D) All of the above

9) When a section of drainage pipe is being tested with water, the section must be tested under a head of water that extends at least _____ feet above the section being tested. Fill in the blank.

A) 4
B) 5
C) 10
D) 12

10) How long is water required to be kept in a section of drainage piping for test purposes?

A) 5 minutes
B) 10 minutes
C) 15 minutes
D) 30 minutes

11) When air is used to test a drainage system, the piping must maintain a pressure of _____ pounds per square inch. Fill in the blank.

A) 3
B) 5
C) 10
D) 15

12) When air is used to test a drainage system with a mercury gauge, the piping must maintain a _____ inch column of mercury. Fill in the blank.

A) 5
B) 10
C) 15
D) 17

13) How long is air required to be kept in a section of drainage piping for test purposes?

A) 5 minutes
B) 10 minutes
C) 15 minutes
D) 30 minutes

14) Before a final inspection is approved, all _____ must be filled with water. Fill in the blank.

A) Pipes
B) Vents
C) Fixture traps
D) All of the above

15) When is a smoke test required for the approval of a final plumbing inspection?

A) In commercial plumbing installations
B) In health-care plumbing installations
C) At the option of the property owner
D) At the option of the local plumbing inspector

16) Building sewers must be tested, when tested with water, with a head of water that is not less than _____ feet above the piping being tested. Fill in the blank.

A) 4
B) 5
C) 10
D) 15

17) Building sewers must maintain a test pressure for at least _____ minutes. Fill in the blank.

A) 5
B) 10
C) 15
D) 20

18) Who is responsible for the maintenance of a plumbing system?

A) The property owner
B) The property owner's authorized agent
C) The plumbing contractor who installed the plumbing
D) Either A or B

19) Plumbing systems must be maintained to remain _____. Fill in the blank.

A) Safe
B) Sanitary
C) Both A and B
D) None of the above

20) Which of the following devices is sometimes required in order to comply with a final plumbing inspection?

A) Ohmmeter
B) Voltometer
C) Manometer
D) Any of the above

I know that 20 questions doesn't seem like much of an exam, but we have covered all the generic aspects of testing plumbing systems. Please answer the more specific true-false questions that pertain to your local area.

True-false test for zone 1

1) The burden of responsibility for the equipment, material, and labor necessary for inspection or tests of a plumbing system shall be that of the person who requests the inspection or test, or that of the person to whom the permit has been issued.

   True False

2) The testing of plumbing systems may be done with air or water.

   True False

3) A water test may be used to test sections of a drainage system, however, water is not allowed as a test media for entire drainage systems.

   True False

4) When a building sewer is to be tested, it first must be plugged at a point not to exceed 5 feet beyond the outside wall of the building foundation where the building sewer is installed.

   True False

5) Twelve hours of prior notice is required to schedule a plumbing inspection.

   True False

6) Requests for plumbing inspections may be made in writing or orally.

   True False

7) A permit holder has up to 10 days to comply with correction or violation notifications made by the code enforcement office.

   True False

8) A plumbing inspector has the authority to require plumbing that was installed and concealed without proper inspection to be exposed again for a suitable inspection.

   True False

9) When water pipes are tested with water, the test water must be potable.

True False

10) When water pipes are tested with water, they must be tested at working pressure, however, when air is used to test the pipes, the test pressure must be at least 40 pounds per square inch.

True False

True-false test for zone 2

1) Rain leaders or downspouts and branches within a building must be tested for leakage with either air or water.

True False

2) Building sewers must be tested with a minimum air pressure of 5 pounds per square foot, when air is used as a test media.

True False

3) A building sewer that is being tested must maintain its test pressure until backfilling of the sewer trench is complete.

True False

4) When water pipes are tested for leakage, they must be tested at a pressure that is at least 25 pounds per square inch higher than their normal working pressure.

True False

5) Peppermint tests are required to obtain any final plumbing inspection.

True False

6) Smoke tests are required to obtain any final plumbing inspection.

True False

7) Master plumbers are required to perform all testing of water, waste, drain, and vent pipes.

True False

8) Reasonable notice must be provided in order to schedule a plumbing inspection.

True False

9) Drainage systems must be tested in their entirety.

True False

10) Tests of water pipes must be maintained for a minimum of 10 minutes.

True False

True-false test for zone 3

1) Plumbing tests pertain to new work, altered plumbing, extended plumbing, and repaired systems.

True False

2) Plumbing that is set up for temporary exhibition purposes is not required to be tested.

True False

3) The replacement of a kitchen faucet requires a code-enforcement inspection.

True False

4) The replacement of a water heater requires a code-enforcement inspection.

True False

5) When water pipes are tested, they must be tested under a pressure equal to, or greater than, their working pressure.

True False

6) Final inspections are made to assure that drainage and vent systems are both watertight and gastight.

True False

7) Existing plumbing is exempt from the maintenance requirements of the current plumbing code.

True False

8) Building sewers must maintain their test pressure for at least 30 minutes.

True False

9) Code-enforcement inspections are not necessary when a master plumber inspects and approves the work of journeyman plumbers.

True False

We have reached the end of the tests on the inspection, test, and maintenance of plumbing systems. We are, in fact, nearing the end of this book. As you know by now, you should check your answers against the correct answers in the back of the book. Compare them, and research any errors you might have made.

I realize this book is not the same as sitting in a classroom and participating in a licensing-exam preparation course. You don't have the benefit of interaction with your instructor and classmates. On the other hand, you don't have the distractions that so often occur when you study with a group.

Answers to multiple-choice exam

1) D	6) C	11) B	16) C
2) C	7) C	12) B	17) C
3) D	8) B	13) C	18) D
4) C	9) C	14) C	19) C
5) C	10) C	15) D	20) C

Answers to true-false test for zone 1

1) T	5) F	8) T
2) T	6) T	9) T
3) F	7) T	10) F
4) F		

Answers to true-false test for zone 2

1) T	5) F	8) T
2) T	6) F	9) F
3) T	7) F	10) F
4) T		

Answers to true-false test for zone 3

1) T	4) T	7) F
2) T	5) T	8) F
3) F	6) T	9) F

CHAPTER 19

Testing your knowledge of engineered plumbing systems

THIS chapter is designed to test your knowledge of engineered or alternative plumbing systems. When you graduate to the status of a master plumber, you are the person responsible for the design and layout of many plumbing systems. There may be times, however, when the design work is done for you. This work is generally performed by engineers, and their work reduces the amount of responsibility you will have to bear.

The plumbing codes in both zones 2 and 3 address the issue of engineered and alternative plumbing designs. The code serving zone 1 is less specific in its handling of engineered systems on an individual basis. All of the codes mandate minimum standards for plumbing installations, but some of them are more detailed in their break-downs of the code.

What do engineered systems mean to you, the master plumber? Well, as long as the layout has been approved by the proper code authorities, all you have to worry about is getting the plumbing installed in accordance with the engineered design. Since the system is drawn by other experts, you don't have to fool around with riser diagrams, pipe sizing, and other aspects of the system design, which can rob you of time and put your name on the line. If you follow the instructions for the design, to the letter, you can't get into water over your head. This is a big advantage, and one which any master plumber should be happy about. Yet, some master plumbers resent working from engineered plans.

Why would anyone dislike working from an engineered plan? After all, the work is all laid out, sized, and ready to install. What could be easier? I'm not sure what it is that makes master plumbers resist engineered plans, but I've certainly known several who scoffed at systems designed by engineers. There might be some jealousy of the engineers' college degrees, of the money they make (although I know many master plumbers who make more money than degreed engineers, doctors, or attorneys), or just simply the fact that the plumbers are not allowed the freedom to do the job their own way. The truth is, if you can work from an engineered plan, your job is easier and you have less risk.

Engineered systems are common in commercial work and rare in residential jobs because of the expense. Most residential plumbers must design the systems themselves. There is often conflict within the plumbing trade as to which type of plumber (commercial or residential) has a tougher job. I've done both types of work, and I believe a plumber who works with residential remodeling work has the most demanding job when it comes to system design and layout. A residential plumber who works in new construction, in my opinion, must have more design skills than a plumber who installs commercial plumbing. My reasoning on this issue is simple. Commercial jobs are drawn, to the last detail, on blueprints for plumbers to follow. Residential jobs rarely offer any type of plumbing drawing on the blueprints, unless the plumber supplies it.

As you have gathered from the previous chapters, a master plumber can be responsible for a great deal of design work, whether it is sizing a storm drainage system or calculating the minimum size of a water service. The design work falls of the shoulders of the master plumber, unless an engineer or design firm is engaged. Regardless of who designs the system, the design must be approved for use by a code official. This chapter is going to help you understand what code officials look for in engineered and alternative plumbing systems. Not only will you gain some valuable knowledge, you will learn facts that you might need to know in order to pass the licensing exam. To get the ball rolling, let's start with a multiple-choice exam.

Multiple-choice exam

1) An alternative engineered plumbing system must conform to the _____ of the plumbing code. Fill in the blank.

   A) Letter
   B) Strict compliance
   C) Intent
   D) None of the above

2) When an alternative engineered plumbing system is installed, it must provide _____. Fill in the blank.

   A) A level of quality equivalent to that required by the plumbing code
   B) A level of fire resistance equivalent to that required by the plumbing code
   C) Both A and B
   D) Either A or B

3) When an alternative engineered plumbing system is installed, it must provide _____. Fill in the blank.

   A) A level of strength equivalent to that required by the plumbing code
   B) A level of effectiveness equivalent to that required by the plumbing code
   C) Both A and B
   D) Either A or B

4) When an alternative engineered plumbing system is installed, it must provide _____. Fill in the blank.

   A) A level of durability equivalent to that required by the plumbing code
   B) A level of complexity equivalent to that required by the plumbing code
   C) Both A and B
   D) Either A or B

5) When an alternative engineered plumbing system is installed, it must provide _____. Fill in the blank.

A) A level of management equivalent to that required by the plumbing code
B) A level of safety equivalent to that required by the plumbing code
C) Both A and B
D) Either A or B

6) When an alternative engineered plumbing system is installed, the _____ must be designed and installed according to the manufacturer's recommendations. Fill in the blank.

A) Blueprints
B) Riser diagrams
C) Equipment
D) None of the above

7) When an alternative engineered plumbing system is installed, the _____ must be designed and installed according to the manufacturer's recommendations. Fill in the blank.

A) Material
B) Riser diagrams
C) Service
D) None of the above

8) When an alternative engineered plumbing system is installed, the _____ must be designed and installed according to the manufacturer's recommendations. Fill in the blank.

A) Grounding
B) Riser diagrams
C) Service
D) None of the above

9) When an alternative engineered plumbing system is installed, the _____ must be designed and installed according to the manufacturer's recommendations. Fill in the blank.

A) Grounding
B) Components
C) Service
D) None of the above

10) The code requires that an alternative engineered plumbing design must be submitted with sufficient technical data to substantiate the proposed plan and to prove that the performance of the proposed system will _____. Fill in the blank.

A) Mirror the language of the plumbing code
B) Work

C) Meet the intent of the plumbing code
D) Not embark on areas not covered in the code book

11) What type of professional is required to submit an alternative engineered plumbing design for review and approval by code officials?

A) Journeyman plumber
B) Master plumber
C) Registered design professional
D) Any of the above

12) What type of person is required to inform the proper authorities that a proposed plumbing system is an alternative engineered system?

A) Journeyman plumber
B) Master plumber
C) Registered design professional
D) The property owner for whom the work will be performed

13) In what manner should code officials be notified of the intent to use an alternative engineered plumbing system?

A) In person
B) By phone
C) By letter
D) On a permit application

14) When an alternative engineered plumbing proposal is submitted to code officials, how many sets of plans, specifications, or documents are required for the submission?

A) 1
B) 2
C) 3
D) 4

15) When an alternative engineered plumbing proposal is submitted to code officials, the documents pertaining to the proposed plans must be _____. Fill in the blank.

A) Signed
B) Sealed
C) Both A and B
D) Neither A nor B

16) The construction documents that must be submitted to code officials for the approval of an alternative engineered plumbing system must include _____. Fill in the blank.

A) Floor plans for the structure being served
B) Riser diagrams
C) Both A and B
D) Neither A nor B

17) Which of the following might be required as a part of a submission for the approval of an alternative engineered plumbing system?

A) Pipe sizes
B) The grade of horizontal piping
C) Both A and B
D) Neither A nor B

18) Which of the following might be required as a part of a submission for the approval of an alternative engineered plumbing system?

A) The direction of flow
B) The grade of horizontal piping
C) Both A and B
D) Neither A nor B

19) Which of the following might be required as a part of a submission for the approval of an alternative engineered plumbing system?

A) The direction of flow
B) The grade of horizontal piping
C) The location of plumbing fixtures
D) All of the above

20) Which of the following might be required as a part of a submission for the approval of an alternative engineered plumbing system?

A) Fixture loads
B) The name of the plumber who will install the plumbing
C) Both A and B
D) Neither A nor B

21) Which of the following might be required as a part of a submission for the approval of an alternative engineered plumbing system?

A) The direction of flow
B) The grade of horizontal piping
C) The location of plumbing appliances
D) All of the above

22) If the documents submitted for the approval of an alternative engineered plumbing system are not approved, who must the code official contact?

A) The property owner of the structure to be served
B) The master plumber
C) The registered design professional
D) The health department

23) If the documents submitted for the approval of an alternative engineered plumbing system are not approved, a code official must contact the appropriate person in which of the following ways?

A) In writing
B) In person
C) By phone
D) Any of the above

24) During the installation of an approved alternative engineered plumbing system, who is responsible for periodic inspections of the installation?

A) The registered design professional
B) A designated plumbing inspector
C) Both A and B
D) Either A or B

25) If discrepancies are found during the installation of an approved alternative engineered plumbing system, who must be notified immediately?

A) The registered design professional
B) The plumbing inspector
C) The plumbing contractor who is installing the system
D) All of the above

26) What is a registered design professional required to do when an alternative engineered system has been installed completely?

A) Sign off on the job
B) Submit a written final report to the code official in charge of the job
C) Consult with the plumbing contractor who installed the system
D) None of the above

27) Local code officials may not issue a notice of approval for an alternative engineered plumbing system until _____. Fill in the blank.

A) The registered design professional requests such an approval
B) The plumbing contractor installing the system requests such an approval
C) A written certification has been submitted
D) All work has been signed off on

28) Who is responsible for stating that a completed alternative engineered plumbing system is in compliance with the construction documents that were submitted to detail the means and methods of the work to be performed?

A) The master plumber for the contracting firm installing the system
B) The local plumbing inspector
C) The registered design professional who designed the system
D) All of the above

29) What is the proper name of a one-way valve that is designed to allow air to enter a plumbing drainage system when negative pressures develop in the piping system?

A) Bottle valve
B) In-line vent
C) Air-admittance valve
D) None of the above

30) What type of valve is intended to act as a substitute for a vent pipe extending to open air?

A) Bottle valve
B) Wilson-vent valve
C) Air-admittance valve
D) None of the above

31) Pipe shall be sized to _____ full-bore flow.

A) Encourage
B) Restrict
C) Prevent
D) None of the above

32) Which of the following types of vents may be allowed to terminate with a connection to an air-admittance valve?

A) Individual vents
B) Branch vents
C) Circuit vents
D) Any of the above

33) What type of device is designed to close by gravity and seal a vent terminal when the internal pressure is equal to, or greater than, atmospheric pressure?

A) A back-flow preventer
B) An airgap
C) An air-admittance valve
D) A well cap

34) When an air-admittance valve is used as a termination point for a circuit vent, the fixtures being served by the circuit vent must be _____. Fill in the blank.

A) Lavatories
B) Clear-water-waste fixtures
C) Of the same type
D) On the same floor level

35) When an air-admittance valve is used as a termination point for an individual vent, the fixture being served by the vent must be _____. Fill in the blank.

A) Connected to a vertical drainage branch
B) Connected to a horizontal drainage branch
C) Connected to storm-water drainage
D) Any of the above

36) Air-admittance valves must be _____. Fill in the blank.

A) New
B) Accessible
C) Unconcealed
D) Installed horizontally

37) What is the minimum number of vent stacks or stack vents that must terminate into open air for any plumbing system?

A) 1
B) 2
C) 3
D) 4

38) Which of the following types of drainage systems may not be equipped with air-admittance valves?

A) Sanitary drainage systems
B) Clear-water waste systems
C) Special-waste systems
D) Nonneutralized special-waste systems

39) When an air-admittance valve is installed, it must be installed in _____. Fill in the blank.

A) A cabinet or recess
B) An attic
C) A ventilated space that allows air to enter the valve
D) Accordance with the electrical code

40) When an air-admittance valve is installed, it must not be installed in _____. Fill in the blank.

A) A space used as a supply air plenum
B) A space used as a return air plenum
C) Both A and B
D) Neither A nor B

41) When an air-admittance valve is installed, the horizontal branch drain that serves the fixture being served by the air-admittance valve must connect to _____. Fill in the blank.

A) A drainage stack
B) A building drain
C) Both A and B
D) Either A or B

42) When an air-admittance valve is installed, the horizontal branch drain that serves the fixture being served by the air-admittance valve must connect to the drainage system at a point that does not exceed _____ branch intervals from the top of the stack. Fill in the blank.

A) 2
B) 3
C) 4
D) 6

43) When an air-admittance valve is installed, the horizontal branch drain that serves the fixture being served by the air-admittance valve must be provided with _____. Fill in the blank.

A) A minimum grade of ½-inch per foot
B) An uninterrupted run
C) A relief vent
D) None of the above

44) An air-admittance valve, when installed, must be installed at least ____ inches above the horizontal branch drain or fixture drain being served. Fill in the blank.

A) 2
B) 3
C) 4
D) 6

45) Air-admittance valves, when installed, must be installed at least _____ inches above insulation materials. Fill in the blank.

A) 2
B) 3
C) 4
D) 6

46) Relief vents that are used in conjunction with horizontal drains and air-admittance valves may _____. Fill in the blank.

A) Terminate into open air
B) Connect to vent stacks
C) Connect to stack vents
D) Any of the above

47) In the selection of drainage pipe, the pipe must be sized to prevent _____. Fill in the blank.

A) Partial flow
B) Full-bore flow
C) Intermittent flow
D) None of the above

48) At what location should a relief vent be connected to a horizontal branch being used in conjunction with an air-admittance valve?

A) Within 3 feet of the fixture being served
B) Within 6 feet of the fixture being served
C) Between the stack or building drain and the most downstream fixture drain connected to the horizontal branch drain
D) Between the stack or building drain and the most upstream fixture drain connected to the horizontal branch drain

49) When dealing with a computerized drainage design, the load must be computed from the _____ conditions from fixtures or the peak usage design condition. Fill in the blank.

A) Simultaneous discharge
B) Sequential discharge
C) Either A or B
D) Neither A nor B

50) Air-admittance valves must be located within the maximum _____ permitted for a vent. Fill in the blank.

A) Developed length
B) Vertical height
C) Horizontal run
D) None of the above

This concludes your testing on the multiple-choice exam. As you know by now, true-false tests follow the multiple-choice exams. These true-false tests are for all three zones of plumbing codes. In this chapter, you will find true-false tests for zones 2 and 3, but none for zone 1 because the code book for zone 1 doesn't have a special section dedicated to alternative engineered systems.

Please proceed to the true-false test for your region, and answer the questions. If you are working in a region covered by the zone-1 code, move on to the answer section in the back of this chapter and check your work on the multiple-choice exam. Keep in mind that you might have trouble with the answers to these questions if you live or work under zone-1 regulations, since the topic of alternative engineered systems is not singled out in that code.

True-false test for zone 2

1) A removable plate, which is commonly attached with screws or bolts, that permits access to pipes and pipe fittings is called an access cover.

True False

2) The lowest point of a trap is called the crown of the trap.

True False

3) Discharge systems are meant to remove foul water from a building quickly and quietly.

True False

4) The function of a fixture trap is to catch foreign objects that might be dropped accidently in the drain.

True False

5) A device that is installed to prevent an exchange of air between the discharge pipe and the building is called a trap.

True False

6) Funnel shaped fixtures reduce the risk of self-siphonage.

True False

7) Both the length and fall of a pipe can be contributing factors in the loss of trap seals.

True False

8) The branch of a trap-seal loss is directly affected by the height of a building.

True False

9) Trap-seal losses associated with stacks can be affected by the flow load.

True False

10) Trap-seal losses associated with stacks can be affected by the height of the stack.

True False

11) Trap-seal losses associated with stacks can be affected by the diameter of the stack.

True False

12) Careful design and installation can produce substantial cost savings in the installation of a plumbing system.

True False

13) All fixtures should be connected to a main stack that has a minimum diameter of 4 inches, where practicable, except in the case of single-story buildings, where a 3-inch stack is satisfactory.

True False

14) When sinks are connected to a separate stack, the stack may be sized smaller than normal.

True False

15) When sinks are connected to a separate stack, the stack must be connected directly to a building drain.

True False

16) When sinks are connected to a separate stack, in buildings with more than 5 stories, the stack diameter must be at least 4 inches.

True False

17) The fixtures on the ground floor in buildings having more than 5 stories must be connected to the building drain separately.

True False

18) Bends and offsets in vertical stacks are prohibited.

True False

19) When 3 sinks are connected to a stack, with the sinks being located over one another on separate floor levels, the minimum diameter of the stack is 2 inches.

True False

20) A 2-story building that is provided with a 4-inch discharge stack may have a water closet, a bath, a basin, and a sink on each floor with no additional venting being required.

True False

21) A 2-story building that is provided with a 4-inch discharge stack may have a 2 water closets, 2 baths, 2 basins, and 2 sinks on each floor with no additional venting being required.

True False

22) When a branch ventilating pipe is required, it must be connected to the individual fixture discharge pipe not less than 6 inches from the crown weir of the trap.

True False

23) A combination waste-and-vent system is limited to sinks, dishwashers, floor sinks, indirect waste receptors, floor drains, and similar devices where the fixtures are not adjacent to walls or partitions.

True False

24) Dishwashers in commercial buildings must drain through grease interceptors.

True False

25) Scullery sinks in commercial buildings must drain through grease interceptors, and they must discharge into a floor sink, through a minimum airgap.

True False

True-false test for zone 3

1) An alternative engineered design for a plumbing system must conform to the intent of the plumbing code.

True False

2) A registered design professional is responsible for submitting technical data that will be used to create an alternative engineered plumbing system.

True False

3) The individual vent air-flow rate for an engineered vent system is determined by a complicated formula.

True False

4) Periodic inspections of the installation of an alternative engineered plumbing system must be performed by a licensed plumbing inspector.

True False

5) Alternative engineered plumbing systems are exempt from the testing procedures required for standard plumbing systems.

True False

6) Air-admittance valves are prohibited.

True False

7) Computerized drainage designs are not allowed for alternative engineered plumbing systems.

True False

8) A 2-inch vent that has an individual vent air-flow rate of 13 cubic feet per minute may have a maximum developed length of 100 feet.

True False

9) Sizing requirements for plumbing drainage systems, when determined by computerized means, must be developed on a commercial software program.

True False

10) At least one vent stack or stack vent is required to terminate into open air for any plumbing system installed.

True False

This concludes all of the tests on engineered systems. Turn to the end of this chapter and check your answers against the correct answers as they are listed. If you are in zone 2 or zone 3, it will be easy for you to determine why an answer is incorrect. Plumbers in zone 1 may have more difficulty determining their needs for engineered systems, since the code book for zone 1 does not offer a detailed section on engineered systems.

Once you have scored your tests in this chapter, you will have completed all of the standard tests and exams. However, you are not quite done yet. Since you are sitting for a licensing exam to become a master plumber, you will have to meet certain math requirements. These requirements are not particularly difficult, but they are an issue with which you must feel comfortable. When you feel confident that you have mastered the chapters to this point in the book, refer to chapter 22 on math, and take the test questions in that section. As with the rest of this book, you will find the correct answers following the last test question. Good luck.

Answers to multiple-choice exam

1) C	14) B	27) C	39) C
2) C	15) C	28) C	40) D
3) C	16) C	29) C	41) D
4) A	17) C	30) C	42) C
5) B	18) C	31) C	43) C
6) C	19) D	32) D	44) C
7) A	20) A	33) C	45) D
8) D	21) D	34) D	46) D
9) B	22) C	35) B	47) B
10) C	23) A	36) B	48) C
11) C	24) D	37) A	49) C
12) C	25) C	38) D	50) A
13) D	26) B		

Answers to true-false test for zone 2

1) T	10) T	18) T
2) F	11) T	19) F
3) T	12) T	20) T
4) F	13) T	21) T
5) T	14) F	22) F
6) F	15) T	23) T
7) T	16) T	24) T
8) F	17) T	25) T
9) T		

Answers to true-false test for zone 3

1) T	5) F	8) T
2) T	6) F	9) F
3) T	7) F	10) T
4) F		

CHAPTER 20

Travel trailers and travel trailer parks

THIS chapter is for students who will be taking the licensing exam in zones 1 or 2. Since zone 3 does not reference the subject of travel trailers and travel trailer parks independently, it is not included in this section. The following questions are all true-false. Answer the questions and then check your work with the correct answers at the end of this chapter.

True-false test for zone 1

1) Recreational vehicle parks are required to provide toilets and urinals in at least one location within the park.

   True False

2) Facilities, housed under the same roof, for males and females must be separated and identified as to which facilities are intended for which sex.

   True False

3) There must be a minimum of 1 toilet for every 10 sites in a recreational vehicle park.

   True False

4) The interior walls in toilet rooms must be moisture resistant to a point at least 4 feet above the finished floor.

   True False

5) Floor drains are required in toilet rooms where flush-type toilets are installed.

True False

6) When designing the fixture layout for a recreational vehicle park, one lavatory must be installed for each toilet installed, until there is a total of 6 toilets; then the percentage changes.

True False

7) Urinals can be substituted for up to one-half of the required toilets in facilities designed for men.

True False

8) Toilets must have open-front seats.

True False

9) Toilet compartments must not be less than 36 inches wide.

True False

10) A minimum water supply of 25 gallons per day, per site, is required for sites that do not have individual water connections.

True False

11) A minimum water supply of 75 gallons per day, per site, is required for sites that have individual water connections.

True False

12) A minimum water supply of 95 gallons per day, per site, is required for sites that have water-flush toilets in provided rest rooms.

True False

13) The minimum allowable water pressure in a recreational vehicle park is 20 pounds per square inch.

True False

14) The minimum allowable water flow in a recreational vehicle park is 2 gallons per minute.

True False

15) The maximum allowable water pressure in a recreational vehicle park is 100 pounds per square inch.

True False

16) Water outlets, when not piped to individual sites, may be up to 500 feet from any site.

True False

17) All potable water connections must be protected against back flow.

True False

18) Potable water connections for recreational vehicles must be installed so that the hose connection is at least 12 inches above the finished grade.

True False

19) Potable water connections for recreational vehicles must be installed so that the hose connection is no more than 16 inches above the finished grade.

True False

20) Up to 5 recreational vehicles can be served by a 3-inch drainage pipe.

True False

21) Up to 24 recreational vehicles can be served by a 4-inch drainage pipe.

True False

22) Up to 71 recreational vehicles can be served by a 5-inch drainage pipe.

True False

23) Up to 110 recreational vehicles can be served by a 6-inch drainage pipe.

True False

24) Up to 400 recreational vehicles can be served by a 8-inch drainage pipe.

True False

25) Cleanouts in the drainage systems for recreational vehicle parks must be installed at intervals that do not exceed 50 feet.

True False

True-false test for zone 2

1) When air is trapped in a drain or drain hose and retards or stops the flow of liquid waste or sewage, the condition is called an air lock.

True False

2) A combination compartment is a compartment that contains a shower stall and a water closet.

True False

3) Combination compartments are sized for occupancy of only one person.

True False

4) A travel trailer that is equipped with a water closet is called a dependent travel trailer.

True False

5) An independent travel trailer is one that is equipped with a water closet and a bath or shower.

True False

6) A service building is a structure that houses only toilet facilities for men and women.

True False

7) Travel trailers are not allowed to be placed in parks where plumbing and sanitation facilities are not installed.

True False

8) Dependent travel trailers must not be parked in a space intended for an independent travel trailer unless a public toilet and bath facility is within 100 feet of the trailer.

True False

9) Before a travel trailer park can provide space for the use of independent travel trailers, the park owner must apply for a permit.

True False

10) Every travel trailer park must have at least 2 service buildings, one for males and one for females.

True False

11) Parks that serve independent travel trailers are only required to provide minimum facilities.

True False

12) Service buildings must be made of permanent construction, with moisture-resistant walls.

True False

13) Service buildings provided for parks that cater only to independent travel trailers are required to have at least one laundry tray, one toilet, one lavatory, and one shower or bathtub for female use.

True False

14) Service buildings provided for parks that cater only to independent travel trailers are required to have at least one laundry tray (for use from both males and females), one toilet, one lavatory, and one shower or bathtub for male use.

True False

15) Hot and cold water must be provided to all plumbing fixtures.

True False

16) The minimum size of a shower stall shall be no less than 32 inches by 32 inches.

True False

17) Every toilet room must be equipped with a floor drain that is 2 inches in diameter.

True False

18) Every laundry room must be equipped with a floor drain that is 3 inches in diameter.

True False

19) All water closets, slop-water closets, tubs, and showers must be in separate compartments.

True False

20) The minimum size of pipe in a travel trailer park drainage system is 4 inches.

True False

21) Main sewers and sewer laterals in travel trailer parks may be installed in a common trench with the park's water service.

True False

22) A travel trailer is considered to consist of 4 fixture units.

True False

23) The mean velocity of flow in a park sewer must be 2 feet per second when the pipe is at full flow.

True False

24) The main sewer must be provided with an individual vent that is not less than 3 inches in diameter.

True False

25) Vents installed on park sewers must be located not more than 10 feet downstream from its upper trap.

True False

26) When long sewer mains are installed in travel trailer parks, they must be provided with relief vents at intervals not to exceed 100 feet, if the manhole covers are not of a perforated type.

True False

27) Sewer laterals serving independent travel trailers must not be smaller than 3 inches in diameter.

True False

28) Sewer inlets must have a minimum diameter of 4 inches.

True False

29) Sewer inlets must extend at least 3 inches above the finished grade.

True False

30) A house trap must be installed for each trailer site.

True False

31) No sewage, waste water, or other effluent is allowed to discharge onto the surface of the ground.

True False

32) Park drainage systems are subject to a static water test.

True False

33) The outlets of branch services of potable water for trailer sites must terminate on the right (as opposed to the left) side of the site.

True False

34) All water services for travel trailers must produce a minimum of 10 gallons of water per minute.

True False

35) A back pressure or reduced pressure principle back-flow preventer must be installed on all water services serving travel trailers.

True False

36) The minimum size of a service connection is ¾-inch pipe.

True False

37) Flexible reinforced hose is an acceptable material for use in connecting the drainage of a travel trailer to a park sewer.

True False

38) The minimum inside diameter for the drain connection between a travel trailer and an park connection is 3 inches.

True False

39) Fittings used for water service connections must be of a quick-disconnect type.

True False

40) Every branch service line for water service must be provided with an individual cut-off valve.

True False

This concludes the tests on travel trailers and travel-trailer parks. Compare your answers with the correct answers at the end of this chapter and research any topics that you answered incorrectly.

Answers to true-false test for zone 1

1) T	10) T	18) T
2) T	11) F	19) F
3) F	12) F	20) T
4) T	13) T	21) F
5) T	14) F	22) T
6) T	15) T	23) F
7) F	16) F	24) F
8) T	17) T	25) F
9) F		

Answers to true-false test for zone 2

1) T	4) F	7) T
2) T	5) T	8) F
3) T	6) F	9) T

10) F	21) F	31) T
11) T	22) F	32) T
12) T	23) T	33) F
13) T	24) F	34) F
14) T	25) F	35) T
15) F	26) F	36) F
16) F	27) F	37) T
17) F	28) T	38) T
18) T	29) T	39) T
19) T	30) T	40) T
20) T		

CHAPTER 21

Manufactured homes and parks

THIS chapter is designed to test your knowledge of manufactured homes and manufactured home parks. The codes from zones 1 and 2 are both represented in this chapter, in the form of true-false tests. The plumbing code for zone 3 does not offer a specific discussion on this subject, so there is no test provided for plumbers in the zone-3 region. You know the drill by now, answer the test questions, and then compare your answers with the correct answers at the end of the chapter.

True-false test for zone 1

1) A mobile home can be a structure of one or more sections.

   True False

2) The body of a mobile home is 8 feet wide, or wider.

   True False

3) To be considered a mobile home, the structure must be at least 30 feet long.

   True False

4) A mobile home is not built on a permanent chassis.

   True False

5) To be considered a mobile home park, a piece of land must be designated and improved to provide a minimum of 4 or more mobile home sites.

   True False

6) When a permit is required to install plumbing in a mobile home park a submittal package must be proposed to the code enforcement office, and the package must contain a plot plan (drawn to scale) of the park that indicates elevations, property lines, driveways, existing or proposed buildings, and the sizes of all mobile home sites.

True False

7) Sewage collection-line systems must be designed for a minimum flow of 150 gallons per day, per mobile home.

True False

8) Each lot for a mobile home must be provided with a 4-inch drain inlet.

True False

9) All lateral lines from drain inlets to sewage drains must maintain a minimum grade of ⅛ of an inch per foot.

True False

10) Sewer inlets may be left open for up to 30 days when a mobile home is removed from its site.

True False

11) The rim of a drain inlet shall not rise more than 6 inches above ground level.

True False

12) Drain inlets must be encased in a concrete slab.

True False

13) The concrete slabs used to encase drain inlets must have a minimum thickness of 4 inches.

True False

14) Drain inlets for mobile home lots must be located within a 4-foot area in the rear third section of the mobile home lot.

True False

15) Drain inlets for mobile home sites do not have to be trapped, so long as the plumbing fixtures in the mobile home are trapped.

True False

16) All drain inlet traps must be vented.

True False

17) The minimum diameter of a vent for a drain inlet is 3 inches.

True False

18) Vents in outdoor locations must extend vertically for a minimum of 10 feet.

True False

19) Wet-vented systems are allowed in the installation of park drainage systems, in lieu of individual vents.

True False

20) When a wet-vented system is used, a terminal vent must be located not more than 20 feet downstream from the uppermost trap on any branch line.

True False

21) When a wet-vented system is used, relief vents must be installed at intervals that do not exceed 100 feet.

True False

22) 3-inch drainage laterals, when installed in a wet-vented system, may be up to 10 feet long.

True False

23) 4-inch drainage laterals, when installed in a wet-vented system, may be up to 15 feet long.

True False

24) When tailpieces are installed with a wet-vented drainage system, they may be up to 30 inches in length.

True False

25) Vertical drainage pipes, excluding tailpieces, may not be installed in a wet-vented system.

True False

26) The drain inlet for a mobile home site should be rated as having 6 fixture units of waste load.

True False

27) The minimum size of a water service outlet for a mobile home lot is ¾ of an inch.

True False

28) Water service outlets for mobile home sites must be located in the same area as the drain inlet for the lot.

True False

29) The minimum allowable water pressure acceptable as a water supply for a mobile home lot is 40 pounds per square inch.

True False

30) Each water service in a mobile home park must have its own cut-off valve.

True False

True-false test for zone 2

1) When a manufactured home park is created, there are several items that must be submitted to code officials before a plumbing permit will be issued. For example, a plot plan (drawn to scale) of the park is required in the proposal package, along with the spaces and area or portion of the park used for setting up manufactured homes.

True False

2) A permit application for installing new plumbing when creating a manufactured home park must include the size, location, and specifications of the park drainage system.

True False

3) A permit application for installing new plumbing when creating a manufactured home park must include the size, location, and specifications of water supply lines.

True False

4) A permit application for installing new plumbing when creating a manufactured home park must include the size, location, and layout of a service building.

True False

5) A permit application for installing new plumbing when creating a manufactured home park must include the size, location, and layout, and specifications for a fire protection system.

True False

6) A permit application for installing new plumbing when creating a manufactured home park must include a layout (drawn to scale) of typical manufactured home sites.

True False

7) Sewers and sewer laterals must be installed in trenches that are not shared by water service piping.

True False

8) Sewers in manufactured home parks must be at least 10 feet away from water services or water distribution systems.

True False

9) 4-inch pipe is the minimum size allowable in the construction of any manufactured home park drainage system.

True False

10) Each manufactured home site will be rated as 15 fixture units when calculating drainage loads.

True False

11) Discharge from a park drainage system must not be connected to a public sewer.

True False

12) Sewer inlets in manufactured home parks must have a minimum diameter of 3 inches.

True False

13) Sewer inlets may extend up to 6 inches above finished grade.

True False

14) House traps are prohibited in drainage systems for manufactured home parks.

True False

15) Sewer laterals that exceed 20 feet in length must be vented.

True False

16) Water service outlets must terminate on the left side of home sites.

True False

17) A water service connection for a home site in a manufactured home park must have a minimum diameter of ¾ of an inch.

True False

18) Laundry trays and washing machines must not be installed in toilet rooms contained in service buildings.

True False

19) At least one 3-inch floor drain is required in each toilet room.

True False

20) At least one 4-inch floor drain is required in each toilet room.

True False

Answers to true-false test for zone 1

1) T
2) T
3) F
4) F
5) F
6) T
7) T
8) F
9) F
10) F
11) F
12) T
13) F
14) T
15) T
16) T
17) F
18) T
19) T
20) F
21) T
22) F
23) T
24) F
25) T
26) T
27) T
28) T
29) F
30) T

Answers to true-false test for zone 2

1) T
2) T
3) T
4) T
5) T
6) T
7) T
8) F
9) T
10) T
11) F
12) F
13) T
14) F
15) F
16) T
17) F
18) T
19) T
20) F

CHAPTER 22

Math for master plumbers

MATH requirements for master plumbers are not very difficult, but some plumbers have trouble with this portion of the licensing exam. Many test facilities will allow you to use non-printing calculators during the test. With the aid of a calculator, the math requirements are much easier. For example, finding the square root of a number with the right calculator is a breeze. Trying to work out the square root of a number on paper, without the aid of an electronic device, is a good deal more difficult.

I am not going to spend a tremendous amount of time on math requirements, but I am going to take a look at some of the basic math skills that you might need to pass the licensing exam. Unlike other sections of this book, this chapter is a free for all. You will get some true-false questions, some multiple-choice questions, and some essay questions. Fasten your seat belt, sharpen your pencil, get some scrap paper, and we will blast off into the math maze for master plumbers.

The math maze

1) What is the square of the number 5?

A) 10
B) 15
C) 20
D) 25

2) What is the square of the number 10?

A) 20
B) 30
C) 40
D) 100

3) Which of the following numbers represents most closely atmospheric pressure at sea level?

A) 12.5 pounds per square inch
B) 14.7 pounds per square inch
C) 19.5 pounds per square inch
D) 27.3 pounds per square inch

4) How many square inches are in 4 square feet?

A) 48
B) 480
C) 576
D) 675

5) How much does one gallon of water weigh?

A) 5 pounds
B) 7.26 pounds
C) 8.34 pounds
D) 9.87 pounds

6) How much does 5 gallons of water weigh?

A) 36.30 pounds
B) 41.70 pounds
C) 49.35 pounds
D) None of the above

7) How many cubic feet are made from 1728 cubic inches?

A) 1
B) 3
C) 5
D) 10

8) If you are installing a sewer with a ⅛ of an inch of fall for a distance of 150 feet, how much lower will the discharge end of the sewer be than the inlet end?

A) 8 feet
B) 16.75 feet
C) 18.75 feet
D) 22 feet

9) How high will a pressure of 100 pounds per square inch raise a column of water?

A) 50 feet
B) 100 feet
C) 150 feet
D) 200 feet

10) How much does one cubic foot of water weigh?

A) 52.5 pounds
B) 62 pounds
C) 62.38 pounds
D) 71.7 pounds

11) How many square feet are in a square yard?

A) 3
B) 6
C) 9
D) 12

12) How many feet are in one mile?

A) 4324
B) 4325
C) 5280
D) 6325

13) How many cubic feet per hour is equal to one gallon per minute?

A) 7.0245
B) 7.5653
C) 8.2080
D) 8.0208

14) If the diameter of a pipe is doubled, what effect will be had on the area of the pipe?

A) It will be doubled
B) It will increase by 2.78 times its original area
C) It will increase by 4 times its original area
D) It will increase by 12 times its original area

15) Assume you are asked to find the square footage of a triangle. This type of situation might arise when calculating a drainage system for roof drains. You are given a triangle with a baseline that is 10 feet long. The height from the baseline to the peak of the triangle is 6 feet. What is the square footage of the triangle?

A) 15 feet
B) 30 feet
C) 60 feet
D) None of the above

16) Quadrilaterals are figures with 4 lines.

True
False

17) A trapezium is one form of a quadrilateral.

True
False

18) If you are given a rectangle with two lines that are 20 feet long and two lines which are 30 feet long, what is the total square footage contained in the rectangle?

A) 60 square feet
B) 150 square feet
C) 300 square feet
D) 600 square feet

19) In dealing with metric measurements, a pipe with a diameter of 76.2 millimeters would be routinely called a _____. Fill in the blank.

A) 1½-inch pipe C) 3-inch pipe
B) 2-inch pipe D) 4-inch pipe

20) Which of the following numbers represents pi?

A) 3.14 C) 3.1416
B) 3.1410 D) 31.416

21) The diameter of a circle can be determined by dividing the circumference by 3.1416.

True False

22) The circumference of a circle can be found by multiplying the diameter by 3.1416.

True False

23) If you multiply the arc of a sector in a circle by the radius of the circle, you will arrive at the area of the sector.

True False

24) A decagon has 8 sides.

True False

25) A parallelogram is a form of polygon.

True False

26) A trapezoid is a polygon.

True False

27) The surface of a cylinder can be figured by multiplying 3.1416 by the diameter and then that number by the length.

True False

28) To find the volume of a solid, you must multiply the area of the base by the perpendicular height.

True False

29) Which of the following numbers is most nearly the square root of 30?

A) 5.4772
B) 7.3432
C) 12.345
D) 12.763

30) If you are reading a set of blueprints that have been drawn to a scale of ¼ inch per foot, how long is a pipe that scales out at 3.5 inches?

A) 3½ feet
B) 9 feet
C) 12 feet
D) 14 feet

31) If you are reading a set of blueprints that have been drawn to a scale of ⅛ inch per foot, how long is a pipe that scales out at 4 inches?

A) 3.2 feet
B) 20 feet
C) 22 feet
D) 32 feet

32) If you are reading a set of blueprints that have been drawn to a scale of ½ inch per foot, how long is a pipe that scales out at 6 inches?

A) 3 feet
B) 6 feet
C) 30 feet
D) 33 feet

33) The formula used to find the cube of the number 2 looks like this: $2 \times 2 \times 2$.

True False

34) The hypotenuse of a triangle is the side that is opposite of the right angle formed by the triangle.

True False

35) The formula used to find the square footage of a rectangle is the same as the one used to find the square footage of a square.

True False

Answers to the math maze

1) D	10) C	19) C	28) T
2) D	11) C	20) C	29) A
3) B	12) C	21) T	30) D
4) C	13) D	22) T	31) D
5) C	14) C	23) T	32) A
6) B	15) B	24) F	33) T
7) A	16) T	25) F	34) T
8) C	17) T	26) F	35) T
9) D	18) D	27) T	

Index

A

B

C

D

J

K

L

M

Q

R

S

T

U